BIRD FROM HELL

FIFTH EDITION

GERALD McISAAC

ISBN: 978-1-957009-23-0 (sc)

Library of Congress Control Number: 2022909783

TABLE OF CONTENTS

INTRODUCTION

Bird From Hell, Fifth Edition

The latest reports from Okanagan Lake are so important, they have forced me to print a new edition of the book, Bird From Hell. Mysteries which have puzzled people for a great many years, are about to be resolved. The animal in Okanagan Lake is the same animal which is in Loch Ness, among other lakes. That animal is Basilosaurus, the walking whale.

As I write this, it is winter. In the spring, this animal will be coming out of the water to graze, in the moonlight. At that time, working people will be able to do that which the scientists have failed to accomplish. They will prove the existence of this animal.

To such people I can only say: I have complete confidence in you!

Chapter 1

Background

The Dene Elders grew up in an environment which is very similar to that which is commonly referred to as the Stone Age. In fact, it was a time of change, of cultural conflict, much as it is now. They were being introduced to civilization, complete with modern tools and weapons, as well as diseases. They were also expected to adapt, to change their social and cultural way of life, in order to fit into civilized society. That same civilized society had no intention of adapting to the life style of Indigenous people.

Those same Dene Elders were members of a hunting – gathering society. It is an understatement to say that life was difficult. All food and tools had to be first gathered or killed. They lived constantly on the edge of starvation. That changed dramatically when the Dene first met traders from civilization. Those traders offered sacks of flour, rice and sugar in exchange for peltries. This is to say that the traders were interested in the hides of animals, such as marten, mink, beaver, lynx, wolf, wolverine, coyote, bear, grizzly and such.

The Dene had their own priorities. They knew how to weave baskets, and used these baskets, especially for gathering berries. These baskets were very useful, but rather feeble. They were not capable of carrying anything much heavier than berries. By contrast, the rice, flour and sugar came in containers of burlap bags. As burlap is very sturdy, the Dene traded peltries for burlap bags. They had no interest in the contents, if only because they did not recognize it as food. The flour and rice was thrown on the ground, while the sugar was thrown into the fire, for entertainment purposes. They enjoyed seeing the sugar sparkle as it burned.

Further, the Dene had no pottery, so the mere act of boiling meat was difficult. The only way this could be done was by using the stomach of an animal which they had killed. Water was added to the stomach, hot rocks from the camp fire were tossed into the water, and after the water started to boil, meat was added. Mountain goat was considered the best for this, as they could get three meals from one container.

This should give some idea of the world in which the Dene lived, and the cultural confusion, which persists to this day. Of course they were nomads, following the herds, unable to stay in one place for any considerable length of time, if only because their food supply was soon exhausted. Their knowledge of the animals in the mountains is without equal. Of necessity, the only way they could survive was by having an intimate knowledge of the behaviour of all the animals. For that reason, I pay strict attention to that which they tell me.

Numerous people have objected to the idea that any group of people would choose to live in such a harsh environment. To such people I can only respond that there was not a great deal of choosing involved.

It is the scientific opinion that there were possibly three migrations of people across the Bering Strait, over a period of thousands of years. The Rocky Mountain Trench provides a natural corridor. The people with whom I live, the Tsay Keh Dene, apparently came across in the latest migration, possibly within the last few hundred years.

These" immigrants" soon found themselves in a situation similar to that which people face today. In other words," not welcome".

The Dene immediately found that the finest" real estate", such as the coastal regions, was already occupied. Such regions provided the inhabitants with a great abundance of seafood, including fish, whale, walrus, seal, sea lion and dolphin. Also fresh water fish, as well as deer, moose, elk, bear, duck, goose and so forth. In addition, the surrounding forests of cedar and hemlock provided the material for lodging and tools. Such coastal areas were as highly prized then, as they are now. Further, the people who lived in those areas were not at all anxious to share their" wealth".

That left the land to the east, the prairies, with vast herds of bison and other wild life, but that too was occupied, and those people too had no intention of sharing. The Dene, as the latest immigrants, had no choice but to subsist, as best they could, in the corridor, the Rocky Mountain Trench.

The Trench is famous for its magnificent scenery and big game. Professional photographers have taken countless pictures of the mountains, while trophy hunters are enthusiastic about killing the biggest male animals, such as moose, elk, caribou and bear. Grizzly and mountain sheep are the top trophies animals, and the hunters pay a small fortune for a chance to kill one of those animals.

The Dene who live in the mountains are more impressed by the difficulty in surviving. The mountains may be truly magnificent, as the poets and photographers can testify. It is also a fact that those same mountains are not the slightest bit forgiving. Any moment of complacency can prove to be fatal. Recently, a couple of girls were reminded of this.

They decided to head for town in the middle of winter, with the temperature close to forty degrees below zero. At that temperature, Fahrenheit and Celsius are almost identical, supremely cold. Their vehicle broke down and they had no two-way radio, so no way to call for help. That and the fact that they had no matches, no way to light a fire, meant that they were in serious trouble. They spent a severely cold night in the pickup and were found the next day. They could not have survived another night in that severe cold.

The only reason they were not rescued immediately was because no one was looking for them. The first vehicle that came along stopped and offered assistance. In the mountains, this is common practice. We tend to take care of each other. After all, the roads in these mountains are not to be confused with highways.

Those two girls lived in these mountains all their lives and had driven those roads many times without any major mishap. As a result of this, they developed a certain complacency, which nearly cost them their lives. Of course they should have carried with them a box of matches and candles, as well as a two-way radio. Under such circumstances, the vehicle provides the shelter and the lit candles provide enough heat to keep body and soul united.

It was only within the last thirty years that roads and bridges have been built, connecting the village with the highway. Electricity is supplied by generators, and the village has running water, a modern school, clinic and store. This has resulted in a vast improvement in the living standard of the people who live here.

The roads leading into the village are maintained by the Forest Service. These are commonly referred to as" dirt roads", while in fact they are gravel roads. They are also sometimes referred to as logging roads, although gravel trucks, ore trucks and low bed trucks also use them. Such vehicles are referred to as trucks, while passenger vehicles are referred to as pickups. Very few cars drive these roads, as they cannot take the punishment.

The roads tend to be rather narrow, with" wide spots" that are suitable for pulling into, when being approached by a" wide load". A loaded logging truck or ore truck, as well as a low bed truck carrying a big machine, qualifies as a" wide load".

At the start of each road, signs are posted. These signs state the name of the road, as well as the correct radio frequency to be used. As well, kilometre signs are posted every kilometre or two. Those who travel these roads are expected to have a two-way radio and announce their location and direction of travel, on a regular basis, perhaps every couple of kilometres. Vehicles which are" loaded", which generally means going in the direction of town, have the right of way. It is up to vehicles which are" empty", or going in the opposite direction, to get out of their way.

Until quite recently, the various" Main Lines", or main logging roads, were assigned certain radio frequencies, and these frequencies were assigned various names. As these names and frequencies varied across the province, it led to a certain amount of confusion. When driving these roads, the last thing anyone needs is confusion. That can cause someone their life. For that reason, the Forest Service decided to change the system, making it more standard. Now each radio frequency is merely given a number, and these numbers are used throughout the province. This simplifies the matter, but also meant that everyone had to buy a new radio. It is a small price to pay for safety. As well, the words loaded and empty have been replaced with" down" and" up".

As an example, a logging truck travelling south to the mill, with a load of logs on the Finlay Main Line, may see a sign that reads "18km". All drivers know that km is short for kilometres, so the driver is expected to get on the two-way radio and announce "eighteen down on the Finlay". If the driver is in a pickup, then that should also be specified. Any vehicles going in the other direction, "up vehicles", are expected to pull over into a wide spot, to allow the down vehicle to pass. It should be noted that people who use these roads tend to refer to kilometres as "clicks".

As for those who consider this to be somewhat unfair, that a loaded truck has the right of way, consider the fact that each logging truck may carry fifty tons of logs. If the truck makes a sudden stop, then the logs it is carrying tend to come forward into the cab of the vehicle. Such things do happen, and they tend to ruin the whole day of the driver.

The side roads off the various main lines lead to logging areas, which we refer to as" logging blocks". After the timber is harvested, the roads inside that block are" deactivated". This means that a machine, usually a backhoe, goes into the area and tears up the road at regular intervals. As well, most of the bridges may be removed. This usually happens only after the block has been replanted.

The road between Tsay Keh Dene and Kwadacha is called the Russel Main Line. It follows the Finlay River, but on the west side of the river. On the east side of the river, there is another road called the Finlay Main Line. Traffic on the Russel Main Line uses a different radio frequency than traffic on the Finlay Main Line. In each case, the vehicles are expected to specify which road they are travelling.

This is in the interests of safety, as we all make mistakes. "Down" vehicles are especially expected to announce their location, while "up" vehicles may also announce, if only less frequently.

No doubt there are a great many people who have a difficult time imagining life in this remote area. Those who first arrive, especially teachers, tell me that the culture shock is severe. It is similar to moving half way around the world. The point being that if the reader has a difficult time in imagining life in these mountains, that is completely understandable. It is for that reason I am going into such details. These mountains are rugged and remote, and are home to several species that are thought to be extinct. They are also rarely seen, if only because there are so few people living in these mountains.

Many years ago, I became convinced of the existence of these animals, and decided to devote my life to proving this. My method is to separate fact from belief. I focus on the descriptions of the various animals, which is detailed and accurate, while setting aside the beliefs of the people providing these descriptions. I respect the beliefs of all common people, members of the public, while not necessarily sharing those beliefs. I have my own beliefs. In turn, I ask all common people to respect my beliefs.

This brings me to the scientists. They are certainly not common people, members of the public. On the contrary, they are highly trained professionals. They too have presented various facts, which are not to be disputed. Bones and fossilized remains speak for themselves. From these facts the scientists have drawn certain conclusions, which they believe to be correct. These conclusions, or beliefs, are referred to as theories. These theories are just that, *beliefs, which are presented as facts!*

Scientific theories are meant to be challenged! That is the one and only way in which we learn about the world in which we live! These theories should not be confused with the personal beliefs of common people, which are to be respected and not challenged. On the contrary, I can only stress that scientific theories are meant to be challenged. The scientists who entertain those theories should not consider these challenges as a personal attack, but merely as a proper application of the scientific method.

Consider the fact that until quite recently, it was thought that we lived at the centre of the universe. Everyone agreed on this, and some people still believe this. It was thought, and some people still think, that all the answers are contained in the bible. That is of course perfectly acceptable, at least for common people.

Such an attitude is perfectly *not acceptable* for scientists. This is not to say that scientists maintain that all scientific questions are answered in the holy bible. On the contrary, they have their own books of science, which they tend to regard as "holy scripture". At least, they treat them as such. The theories put forward in certain science books are *not* to be challenged. Any scientist who challenges those theories faces career suicide. As a result of this, students of science have learned to memorize scientific theories, rather than challenging those same theories.

Many years ago, it was scientists such as Kepler, Copernicus and Galileo who devoted their lives to science. They first challenged the theory of the earth being at the centre of the universe. In fact, they risked their lives in studying the night skies. If they had been caught, they would have been executed. It was Newton who studied the work of those scientific pioneers and brought it all together in his three laws of motion. As Newton put it, he had merely "stood upon the shoulders of giants".

Since the time of Newton, the state of science has regressed. Once again, scientific theories are meant to be memorized, not challenged. Anyone who dares challenge scientific theories is not allowed to earn a living in any field of science. For that reason, I am appealing to the members of the public. Working together, we can make a number of major scientific breakthroughs, if only by proving the existence of these animals.

I first became convinced of the existence of these animals when my mother in law recognized a picture of an elephant. She mentioned to my wife that the elephant she remembered was "hairy". For that reason the Dene refer to it as the "hairy elephant". Of course that is a reference to the woolly mammoth, and more on that animal later.

The main thing is that my mother in law lived almost all her life in a log cabin. She spoke broken English, gave birth to all of her children in the mountains, and only after the birth of all of her children, did she first set foot in the place she had heard so much about. That place is called "town". That is the same person who recognized a picture of an elephant! The only explanation is that she saw a woolly mammoth in her younger years! With that in mind, I began to seriously consider the fact that the stories they were telling were completely true.

CHAPTER 2

THE DEVIL BIRD

If there is one thing which terrifies the Dene Elders, it is the animal which they refer to as the devil bird. They are convinced that it is an evil spirit, a demon, a bird from hell. As bullets have no effect on spirits, their fears are well grounded. This is not to say that I share their belief that this animal is a spirit, but it is to say that I respect their belief.

Numerous people are aware that the Indigenous Elders all across North America, and not just the Dene, believe in spirits. They also know that it is the belief of such Elders that these spirits fly around in the darkness and hunt people. Yet as far as I am aware, no one has ever asked the Elders to describe these spirits. That did not stop me from asking the question, and the answer I received was a complete shock. They told me that "it has a head like an eagle, the body is the size of a mans body, the wings are as wide as two moose hides stretched out, the flapping of the wings sounds like dry hide, and the tail is as long as a man is tall and it ends in the shape of an arrowhead."

I could not possibly have been more surprised, as they provided me with a near perfect, detailed description of a pterodactyl! More accurately, they described a flying reptile which is technically referred to as pterosaurs. The one with the long tail is classified as rhamphorynchus and is more common than the animal with a short tail, which is classified as the true pterodactyl. The wing span of these flying prehistoric monsters is estimated to be between ten and fifteen metres, or thirty five to fifty feet. These animals are the size of small planes! They are also strong enough to pick up deer and pack them away! As deer weigh possibly ninety kilos or two hundred pounds, that is quite an accomplishment. No wonder the Elders are afraid of it!

Allow me to state as a fact that the Dene Elders do *not* go outside after sundown. They maintain that "a big bird" will "pick you up and carry you away". They have pointed out to me certain caves in the mountains. These caves open up onto a vertical face. It is their belief that those caves lead directly to hell. These beliefs I respect. Then after sun down they hear a snapping sound, which they think is the sound that Satan makes when he opens the gates of hell. I think it is the sound the animal makes when it opens its wings. They also tell me that after it comes out of the caves, it climbs up the vertical face to the top of the mountain.

Without doubt, the animal is nocturnal, which is to say that it comes out of those caves only after sun down, and returns to the caves before sun rise. It avoids the light, and not only sun light. It avoids all light, including the artificial lights of towns and cities. That is the reason it is so rarely seen. This also makes it more difficult to detect. But then there is more than one way to skin a cat!

The caves in those mountains are the nesting grounds of the flying reptiles, referred to as pterosaurs or pterodactyls. The mountains are quite distinctive, as the mouth of the caves open up onto a vertical face, commonly referred to as a cliff or a bluff. The top of these mountains are flat or horizontal, so with the combination of vertical and horizontal, I refer to these mountains as perpendicular mountains. My point is that the simplest, easiest way to locate these animals is to quit looking for them and instead, look for perpendicular mountains. When we find such mountains, complete with caves, we will find the nesting ground of the pterosaurs. I refer to this approach as the indirect approach, and I swear by this. There is no point in attempting that which has been attempted countless times before, without success. The medical professionals tell us that to repeat that which has been tried on countless previous occasions, while expecting a different outcome, is the very definition of insanity.

It is important to remember that the mouth of the caves must be only high enough that predators, such as bears and wolves, cannot climb into the nest. The mountains which contain the nests of these flying reptiles does not have to be huge. As long as the entrance is four or five metres, fifteen or twenty feet above the ground, that is sufficient.

It has also been pointed out that as these animals are so huge, the size of small planes, is it possible that they can be seen on radar? To this question I can only respond that the question is backwards. The real question should be, how can they *not* be seen on radar?

These animals are reptiles, which reproduce by laying eggs. Further, reptiles cannot generate their own body heat. The eggs of all reptiles have to receive heat from an outside source, as otherwise they cannot hatch. Bear in mind that the temperature inside those caves is perhaps 13 degrees Celsius or 57 degrees Fahrenheit. That is far too cool for eggs to hatch, even the eggs of reptiles. The eggs of reptiles are able to hatch at a lower temperature than those of birds. In fact, the eggs of lizards are able to hatch at a temperature of around 70 degrees Fahrenheit or 20 degrees Celsius. Fortunately for those flying reptiles, the day time temperature in early May, at least in this part of the world, frequently reaches that temperature.

For that reason, in early May, the females are forced to come out of the caves, build a nest and lay their eggs. Their instinct to reproduce becomes greater than their instinct to avoid the light. They choose a hill top which is reasonably bear, open to the sun, and not too high. They then stay outside the caves and guard the eggs. When the temperature drops during the night, they sit on the eggs. Their bodies act as insulators, to ensure the heat that the eggs have absorbed during the day time, stays within the eggs. During the day, the animal guards its nest by lying down, close to the nest. It is then able to change its skin colour, in order to blend into the background, so that it resembles a lump on the ground.

I should add that the place where the female builds a nest and lays her eggs, I refer to as the nesting *site* of the pterosaur. I use the word site to distinguish it from the nesting *ground* of the animal, which is the caves in the mountains. The nesting site varies from year to year, as every predator in North America is constantly looking for those eggs, and bare hill tops do not stay bare for long.

The eggs are the size of ostrich eggs, according to one of my buddies. He once found such an egg, and said that it had a "thick skin". At the time, he had no way of knowing that the eggs of all reptiles have thick shells. It is simply characteristic of reptiles. He also had no way of knowing that the egg which he had stumbled upon was worth a fortune. At least, it was worth a fortune until someone destroyed it!

The eggs of ostriches take forty two days to hatch. It is reasonable to assume that the eggs of these pterosaurs require a similar time to hatch. Assuming that to be the case, then the eggs hatch around mid to late June. At the time of the hatching of the eggs, the females are placed upon the horns of a dilemma. *Reptiles do not feed their young!* Once the eggs hatch, then the little guys are on their own. They have got to fend for themselves. The trouble is that the female builds the nest on hill tops which are bare and open to the sun. This provides the eggs with the heat from the sun, which they need in order to hatch. It also provides for a breeze which serves to drive away the bugs. It is these bugs which are the very thing the youngsters need in order to survive.

I should mention that newly hatched birds are referred to as chicks. By contrast, newly hatched flying reptiles are referred to as flaplings. The point being that as soon as the eggs begin to hatch, the female is forced to pick up the nest and carry it to a place where there is a great deal of food for her youngsters. That place is a swamp. The food is mainly bugs, of all varieties, although I am sure they tend to feast mainly upon mosquitoes. Within possibly six weeks, the youngsters are strong enough to fly. I know this for a fact because the members of the village reported to me that they saw them flying at the end of July.

The Dene, as well as all Indigenous people, are well aware of the existence of these flaplings. As the eggs from which they hatch are the size of ostrich eggs, the freshly hatched youngsters are the size of ostrich chicks. Further, the wings are not fully developed at the time of hatching, so the flaplings are forced to run around on two legs, devouring as many bugs and very small animals as they can.

At the same time, it is a matter of avoiding predators, not an easy task. For that reason, very few flaplings live long enough to learn how to fly.

It is entirely possible that the female stays close to her flaplings, guarding them as best she can, until they are old enough to fly. The reason I say this is because crocodile females guard their youngsters for some time, after they hatch. In this way, perhaps one percent of the young crocodiles live to become adults. Perhaps the female pterosaur also guards her flaplings, until they are old enough to fly, and then leads the few surviving youngsters to the caves in the mountains. As yet, we have no way of knowing.

In regards to the scientific community, the fact is that they have discovered the fossilized trackways of the pterosaurs. Those tracks tell us that as adults, the animals sometimes walk on two legs and sometimes they walk on all four limbs. In other words, the wings double as legs. That clarifies a few things. It explains the existence of the animal the Dene refer to as the "devil dog", which is nothing other than the pterosaur walking on all four limbs. The stories of the "devil bird" is a reference to the pterosaur walking on two legs. It also explains how the animal can climb up a vertical face. It climbs up the cliff on all fours.

I am using this as an example of separating fact from belief, as it applies to scientists. The facts they present are accurate. Their belief that these animals are extinct is not based upon any scientific basis. That belief is merely a theory which is presented as a fact, which is completely contrary to proper scientific procedure.

To return to the subject of the flaplings, the Dene have a name for these animals. In their own language, they call them "little people". In other part of the world, these animals have other local names. The most popular local name is "leprechaun".

The point is that these flaplings, or "little people", or "leprechauns" will be seen only in early summer, and only in or around swamps.

As I have a twisted sense of humour, I once made the mistake of mentioning to a couple of girls, that as they objected to being referred to as "chicks", they should count their blessings. They should perhaps be grateful that they are not called "flaplings" or "leprechauns". I even performed my finest W.C. Fields imitation, of "my little leprechaun". They responded by letting me know what they thought of me and my little joke, in words which do not bear repeating. But then not everyone appreciates my sense of humour.

On a more serious note, other questions came to mind, and bothered me until I did a little research and came up with some answers. One such question is, why are these animals nocturnal?

A little research reveals that these flying reptiles, pterosaurs, which are thought to be extinct, evolved around two hundred forty million years ago, on a huge landmass, referred to as pangea. The scientists refer to this landmass as a "super continent", which it was not. It was the grouping together of all seven continents of the world, so that it was not one continent but seven, and there was nothing "super" about it. That is the reason I refer to it as the "world landmass" of pangea. The flying reptiles, pterosaurs, then spread around pangea, which is to say that they spread around the world. More accurately, they spread over the seven continents. Over the millions of years since, they may or may not have spread to various islands, such as Hawaii and Tahiti.

Over a period of many millions of years, the world land mass of pangea split up, and the continents went their separate ways. Each continent carried the pterosaurs with them. As the continent of Antarctic drifted ever closer to the south pole, it gradually became too cold for the eggs of these reptiles to hatch. For that reason, the pterosaurs died out in the Antarctic. *The pterosaurs still exist on the other six continents!*

I can only stress that these flying reptiles are very much alive in North America, South America, Asia, Africa, Europe and Australia. Further, these animals are predators. They hunt in darkness, in open areas. As predators, they have a keen sense of smell and are far more likely to attack when they smell blood. For that reason, they frequently prey upon girls of child bearing age. They also prey upon children, because children are easier to pick up and carry away.

No doubt, at the time these flying reptiles first evolved, the males had head crests and large beaks. These served as "decorations", adornments to impress the females during mating season, although it is possible that the long beaks also served some practical function, such as catching fish. It is very likely these animals were most active in day light, as they had no competition, aside from each other. They had no particular reason for avoiding the light of day.

That changed dramatically at the time the flying birds first made an appearance. Some scientists maintain that the animal called archaeopteryx was the first flying bird, and it evolved around 150 million years ago, or MYA. Other scientists most emphatically disagree, arguing that it could not possibly have taken flight. It is a subject upon which I have no opinion. The important thing is that flying birds evolved, many millions of years ago, and some of those flying birds became raptors, birds of prey. Of equal importance is the fact that those raptors challenged the flying reptiles for control of the sky.

This brings us to an item of great confusion, which is of course "dinosaurs". The word literally means "terrible lizard", and first found it way into the scientific literature many years ago. It has since become a household name, and is deeply entrenched. That is most unfortunate, as the name is completely inaccurate. For that reason, I try to avoid it, whenever possible. It is not always possible.

The misunderstanding began about two hundred years ago, when scientists of the day discovered fossilized remains of huge animals. They were at a loss to explain the existence of these animals, as they could find no reference to them in the bible. At that time, the bible was the only reference source the scientists had available. This was in the days before Darwin. So they did the best they could and called these extinct animals "dinosaurs", or "terrible lizards".

I have made my position quite clear in other writings. I maintain that the animals which are commonly referred to as dinosaurs are nothing other than birds. They had feathers and they laid eggs. Certain species died out, which happens, while other species evolved and gave rise to new species, which also happens. *There was no mass extinction of dinosaurs!*

Almost all scientists maintain that there was a mass extinction of dinosaurs, and argue passionately concerning the cause of that extinction. All are mistaken, and many have argued that the flying reptiles were dinosaurs. Most scientists also maintain that there has never been a mass extinction of reptiles, while maintaining that no less than five orders of reptile have gone extinct. Of course the public is confused!

To return to the time, many millions of years ago, when raptors first challenged the flying reptiles for control of the airways. The war was immediate and brutal. No quarter! Strictly a matter of survival of the fittest. There was no middle ground. The results were clear for all to see. As anyone who has ever eaten a traditional Christmas dinner can testify, we feast on roast turkey, a bird, and not roast pterosaur, a reptile. Clearly, the raptors won the war for control of the sky, or at least during the day time.

How is it that the raptors were able to drive the flying reptiles out of the day time sky, into the relative safety of the darkness? Was their flying ability superior to that of reptiles? To answer that question, we must examine the flying skills of birds, reptiles and bats.

We know that in order for a body to become airborne, there must be a lifting force on the wings that offsets the downward force of gravity. That lifting force is created when a wing is curved on the top and flat on the bottom. As the animal flaps its wings, the air that travels over the curves must travel faster that the air below the wings. This difference in air speed creates an area of low pressure above the wings. The air from below the wings pushes up on the wings, which creates lift. Of course, the wings of birds have a curve on the leading edge. The flapping of the wings creates lift, and the animal becomes airborne.

Modern aircraft operate on the same principle. Flaps on the leading edge combat turbulence, which can be a problem when air speed is low and the aircraft is trying to land.

We can first examine the flight of bats. These mammals are very efficient fliers, and for good reason. The hind quarters of a bat, commonly referred to as its legs, are attached to its wings. The flapping of the wings is powered not only by the front quarters, called the arms, but also by the legs. When the bat flaps its wings, the legs move up and down with the arms. The result is a very strong wing stroke, which results in greater lift. The tracks left by the bat are also distinctive as they walk on the front feet as well as the hind feet, with the hind feet splayed out far to the side.

The recently discovered fossilized trackways of the pterosaurs show that the animal walks in a manner similar to that of a bat- sometimes on two legs and sometimes on all fours with the hind paws spread out, clear evidence that the hind paws are attached to the wings. All four limbs of the pterosaurs are involved in the flapping of the wings. Clearly the pterosaur is a most efficient flier, far more so that I expected. Almost all the muscles of the body are involved in the flapping of the wings, which is to say flight. It is also a fact that this animal has a special plate of light weight bone that strengthens the torso and shoulders, thus eliminating most of the muscle groups that do not contribute to flight. This gives it an advantage. The more light weight the flying animal, the easier it is for the animal to become airborne and stay airborne. The pterosaur has flaps on the leading edge of the wings to combat turbulence, thus making it more stable, just as modern aircraft have flaps. Even this is not the only flying advantage of the pterosaur. The shoulders of the animal are a marvel of aerodynamic engineering. It is very likely that this animal has an extra pivot joint between the upper end of the shoulder blade and the bony plate that stiffens the torso, which allows the shoulder to swivel. This permits the shoulder bone to swing not only up and back, but also down and forward. Such movements dramatically increase the power of the down stroke of the wing.

In addition, as cold blooded animals, reptiles need to consume far less food than birds, which are warm blooded animals. These are all to the advantage of flying reptiles.

We can contrast these advantages to that of birds, starting with the flight feathers. Such feathers are attached to the wings and only to the wings. Only the front quarters of birds are involved in flapping the wings. The hind legs of birds are dead weight when the animal is flying.

On the other hand, as warm blooded animals, birds do not require the heat from the sun to warm up and become active. At the first crack of daylight, birds, including raptors, took to the air and began to hunt – as they do to this day. By contrast, pterosaurs, as flying reptiles, had to wait for the heat from the sun to warm them up before they could become airborne. The amount of time required to warm up the reptile varied, depending upon the size of the reptile, the night temperatures, the weather conditions, the time of year and numerous other variables.

The brief interval between dawn and the time that the heat of the sun warmed up the flying reptiles was critical. Although the interval may not have been long, it was long enough for the predatory birds, the raptors, to decimate the flying reptiles, the pterosaurs. Even though the cold blooded reptiles were far more efficient fliers and required far less food, pound for pound, the raptors still out competed them for mastery of the daytime skies. It is very likely that over a period of tens of millions of years, various species of raptor evolved and gradually drove all species of flying reptile, pterosaurs, either to extinction, or to the relative safety of darkness. This is another way of saying that the flying reptiles became nocturnal, and this happened around the world.

This is *not* to say that the raptors have won the war, and there is no more competition between birds of prey and flying reptiles. It just means that the battle ground has changed, from the day time skies to the night time skies. Raptors such as owls and night hawks hunt in the darkness, as do bats, in competition with flying reptiles, pterosaurs.

As the pterosaurs gradually adapted to a life of living in darkness, they were forced to make a few adjustments. In particular, during mating season, the females were no longer deeply impressed by the long beaks and head crests of the males, if only because they could not see them. The males had to find another way to display, and they did. The males evolved the ability to glow.

As for those who find this quite remarkable, consider the fact that "fireflies" have evolved the ability to glow. These insects, otherwise known as glow worms or lightning bugs, are really beetles. Light production in these beetles is due to a chemical reaction called bioluminescence. The beetle has a special light emitting organ in its lower abdomen, just as there is a no doubt a similar light emitting organ in the pterosaur.

In the case of the firefly, there is an enzyme, luciferase, which acts on the luciferin in the presence of magnesium ions, ATP, and oxygen to produce light. The light is in the visible spectrum of 520 to 680 nanometers so we see the light as yellow, green or pale red.

No doubt a similar chemical reaction occurs in the pterosaur, which gives off a glow from the torso and is clearly visible from the ground. As yet, we do not know the precise wavelength of light, which is to say the colour, of the light produced. We know that each species glows in a distinctive colour. These colours are as distinctive as the plumage on birds, and will soon be used to help identify the species of pterosaur.

It is to be hoped that many readers will be inspired to look into the chemistry involved in the production of such bioluminescent light.

It is very likely that the males acquire the ability to glow only at the onset of sexual maturity. As yet, we have no way of knowing if the females also have the ability to glow.

The reason I say this is that there are times when evolution will take one characteristic and use it for more than one purpose. Herds of caribou come to mind. Among such animals, both males and females grow antlers. The males grow antlers in the spring and summer in order to display in the fall, and then drop the antlers in the winter. But then the females grow antlers in the fall and winter in order to protect their calves which are born in the spring, and then drop the antlers in the summer.

In much the same manner, among pterosaurs, the males use the ability to glow, in mating season, as a means of display. But then those same animals also use the glow for hunting purposes! They will fly around a huge swamp glowing brightly. The bright light attracts clouds of insects, and those insects attracts bats. Then the flying reptiles attack the flying mammals.

As yet we have no way of knowing if the females also glow, under these circumstances. We do know that many people find these lights to be quite entertaining, and in certain communities, are used as a tourist attraction.

We also know that this glow from the torso of the pterosaurs is the basis of a great many Unidentified Flying Objects, or UFOs. As I have now identified many of these flying lights, it is no longer accurate to refer to them as UFOs but instead as IFOs, Identified Flying Objects. Of course, that is not about to happen, but hopefully, it will appease many of my most vocal critics who have been complaining that I have not been looking into UFOs. God forbid that I should be accused of being a UFOlogist! For many year it has been my opinion that such people are not entirely sane, to put it politely. More accurately, I have long maintained that they are completely out of their minds. I can only hope that they will adopt a more scientific approach to their research.

These lights in the night sky, so called UFO's, have been associated with the death and mutilation of livestock. A popular television show has recently stumbled upon the fact that in the mountains around the world, horses and cattle which are left outside after sundown are being discovered in the morning, "dead and mutilated". The fact that there is no blood around the wound sites has everyone puzzled. Further, this death of livestock has been associated with lights in the night sky.

The writers of the television show have noticed the connection between the UFO's and the dead livestock. They have come up with their own hare brained theory to explain this.

The "explanation", as put forward by the announcer, is that aliens from a distant galaxy have travelled to planet earth, presumably at "warp speed", what ever warp speed is, have "beamed" these animals up to their spacecraft, drawn out their blood and mutilated them, and then "beamed" them back down to planet earth. As "proof" of the existence of these spacecrafts, the commentator has mentioned the existence of these lights in the night sky and the motion of these lights, motion which does not match the motion of any mechanical aircraft, either plane or helicopter. Of course the reason for this is quite simple. The flight of these reptiles is not mechanical!

As this television show is so popular, it is safe to conclude that such nonsense is considered to be entertaining. It is further safe to conclude that the level of scientific education in North America has been reduced to an absurdity.

It is to the credit of the television show that they have drawn attention to the fact that livestock is being killed "around the world", as they state it. This is not entirely true, but then the writers of the show are not focused on accuracy, but on entertainment. In fact, it would be more accurate to say that livestock which is located close to mountains are being killed around the world. Further, it is very likely that the "UFOs" are seen only in the spring, during mating season, but that too is of no interest to the entertainers.

This in no way changes the fact that livestock which is left outside after sundown is being discovered in the mornings, dead and mutilated. The owners of these animals are completely puzzled, as they have clearly not been shot by humans. Nor have they been killed by predators such as bears, grizzlies, wolves or mountain lions. The fact that there is no blood around the wounds leaves no room for any misunderstanding on this point!

This lack of blood around the wound sites is of critical importance. It tells us that the animal was dead when it was mutilated. The question then becomes, what killed the animal? We can rule out blood loss, so that leaves poison. What kind of poison? It certainly did not eat the poison, so it must have inhaled the poison. As that is the case, the poison gas which killed the animal must be within the blood which is within the carcass. If a sample of this blood can be withdrawn within twenty four hours of the death of the animal and sent to a laboratory, then it is very likely that the lab will be able to determine the poison gas which was used to kill the animal.

The livestock was clearly killed by a pterosaur, otherwise known as a pterodactyl. There are various local names in North America, such as devil bird or thunder bird. In other parts of the world, such as Asia and Africa, they are commonly referred to as dragons or fire breathing dragons. The reason they are referred to as fire breathing dragons is because of the "cloud of smoke" they frequently release. Except that this "smoke" is not smoke at all, but toxic. In fact it is so toxic, it is capable of killing a horse. We know this for a fact, because that is precisely the thing that kills this livestock.

As the subject is so important, I will mention that it is characteristic of the animal to rip out the genitals, as well as other "morsels" such as the eyes, nose and tongue. It also frequently rips out the large intestine, from the rectal orifice. As the large intestine is very nourishing, this makes complete sense. The animal is not strong enough to rip apart the carcass so it has to settle for the items it can easily gather.

The standard response of the farmer is to call the police. As he is well aware that the animal was not killed by a predator, or at least not by any predator of which he is familiar, he just naturally assumes that it was killed by a human. Such an assumption is natural, but mistaken. As the police respond, they too are puzzled, as it was clearly not killed by any human. They note the lack of bullet wounds, foot prints, tire tracks or blood. This lack of blood is key.

It is reasonable to assume that while the pterosaur is consuming the carcass, it is also leaving behind some trace DNA. With that in mind, I can only suggest that a swab be used to wipe around the edge of the wound sites and also sent to a lab, along with a blood sample. A laboratory analysis could probably determine that the DNA is that of a reptile. It is not likely that we have the DNA of a pterosaur with which to compare it, but just to determine that the animal was killed by a reptile, would be a step in the right direction.

It is vitally important to prove the existence of this animal. The members of the public have got to be made aware of the fact that it exists, it is a predator, it hunts in darkness, in open areas, it has a keen sense of smell, and is far more likely to attack when it smells blood. This helps to explain the disappearance of so many girls of child bearing age!

The highway from Prince Rupert, on the West Coast, to the city of Edmonton, in the central interior of the province of Alberta, is known as the Highway of Tears. It has earned this name, over the years, because so many people have disappeared from this stretch of highway. It is not by chance that this highway runs directly through the mountains! Nor is it by chance that most of these people have been girls of child bearing age!

The sad fact is that families break up, and when that happens, all too often, young girls leave home. In the cities and town, they generally jump on a bus. Human predators are well aware of this, and consistently hunt for these girls in bus depots. By contrast, the girls of families who break up in isolated areas, including many Reserves, tend to leave home and go hitch hiking. These girls are then preyed upon by the pterosaurs. They generally attack because they smell the blood!

Of course, it is not just on highways that people are attacked by these flying reptiles. They hunt in all open areas, in the darkness. The areas have to be open because the animal has a huge wing span, and the area has to be dark because the animal tends to avoid the light. On the other hand, there are indications that the animal is becoming "braver", more familiar with artificial lights, so that such lights do not offer full protection, as was once the case.

After the animal comes out of caves, in the darkness it generally selects an evergreen tree and with its paws, pushes over the top stem of the tree, frequently a pine or spruce. Then, when the stem of the tree is horizontal or "flat", to phrase it in common parlance, the animal sits on this perch. These trees are usually located at the edge of a clearing, as the animal hunts in open areas. These clearings may or may not be on the edge of a highway or road. The animal also hunts near camp grounds, play grounds, school yards and in fact near any open area, and that includes the backyards of people. They tend to return to the same tree night after night. As a result of this, the top stem of these trees tends to grow on an angle, or a "slant", as is commonly stated. To identify these trees is to identify the hunting spot of the pterosaur.

As it is only the top 4 to 5 meters or 12 to 15 feet of the tree which is bent over in this manner, that provides us with a "ballpark" figure as to the weight of the animal. I estimate the animal weighs no more than 10 to 15 kilos, or 25 to 30 pounds, 40 pounds at most. I refer to this as an example of using "common sense", and it is one of the methods I recommend. I do not view this as an alternative to the scientific method, but as part of the scientific method.

I should add the fact that as the animal prefers to sit on the top of evergreen trees, this helps to explain the reason that experienced woodsmen prefer to build their cabins within clearings. It is partly to avoid the chance of a strong wind blowing the tree onto the cabin, but also because such woodsmen are aware of the existence of this animal, and are also aware that it likes to sit on the top stems of trees.

In the winter months especially, the woodsmen have to come out of the cabin after dark, if only to go to the backhouse or to the wood pile. If a tall tree is close by, they have no way of knowing if a "devil bird" is sitting there, just waiting for them to step outside.

My suggestion, for those who are caught outside after sundown and are attacked by this animal, is to seek shelter. If no shelter is available, then as a last resort, feel free to "hit the ground". This predator attacks us by wrapping its paws around our shoulders and sinking its claws in. Why make it easy for them? If we are flat on the ground, it is more difficult to kill us.

As for those who think that the paws could not possibly be big enough to wrap around our shoulders, think again. Those who have seen the tracks that these animals leave in fresh fallen snow can testify to that. They assure me that the paws are possibly half a meter long or 18 to 20 inches, and 6 to 8 centimetres or 2 or 3 inches wide. As they phrase it, "real long and real skinny." It is not by coincidence that these descriptions match the fossilized trackways of the pterosaur.

These animals have truly adapted to a nocturnal lifestyle. They come out of those caves only after sundown (aside from the females which emerge in order to build a nest and lay eggs), at a time which is generally cooler than the day time temperatures. In fact, these animals are able to hunt in temperatures as cold as twenty degrees below zero Celsius, or zero degrees Fahrenheit. These reptiles are able to do this only because they are very large. Only the largest of reptiles are able to withstand such cold, and only for short periods of time, due to a process referred to as gigantothermia. It is precisely the same process that allows another species of reptile, the leatherback turtle, to survive in the very cold environment of the North Atlantic.

While outside the cave, the pterosaur loses little heat, partly because a large body size leads to a small surface area to volume ratio, so that the heat exchange volume remains low. Consequently, the core body temperature is slow to change, while a spherical body and a layer of fat helps a great deal.

Local people know the animal is close and on the hunt because they recognize the sounds that it makes. It is able to imitate the various sounds that it hears. Those sounds include the sounds of "dogs barking, babies crying and women screaming". Without doubt, they absolutely hear the sound of women screaming. In fact, each time they attack women, they hear the screams.

No doubt, closer to town, they can also imitate the sound of car horns and train whistles, as they hear them so frequently.

For the sake of completeness I should mention that frequently, at the time of the attack of the animal, it releases a very loud whistling sound.

That is the reason for one of the more common names of this animal, the "thunder bird".

The animal is also reasonably intelligent, at least for a reptile. I was surprised to find that they have learned to associate gun shots with food. The local boys found that out when they were "spring trapping" several years ago. At that time of year, towards March and April, the fur of the land dwelling animals is no longer valuable, as the warmer day time temperatures adversely affects the winter coat of the animal. The fur of water dwelling animals, such as beaver, mink and muskrat, remains in prime condition during that time. For that reason the local trappers hunt them at that time, mainly focusing on beaver. Their method is to set traps for them and sometimes to shoot them when they come out of the water. The boys found that on the days they shot beaver, they could expect company after sun down. On the days there was no shooting, the "devil bird" did not drop by. Clearly the animal has learned to associate gun shots with food.

This brings us to Missing and Murdered Indigenous Women, MMIW, which is thought to be a national disgrace, as indeed it is. Then again, the title is not entirely accurate, because it it not only Indigenous women who are disappearing. Members of other ethnic groups are disappearing, as well as children.

As previously mentioned, most of these people are females of child bearing age, which the animal attacks because it smells the blood. Some of the people who have disappeared are children, as children are easier to pick up and carry away. I mention it again as it is so important. This is not to say that human predators are not responsible for the disappearance of Indigenous women, as well as others. It is just that the disappearance of such people, from open areas, in the darkness, is mainly the result of predation by the pterosaurs.

The members of various police departments, across the country, are under extreme pressure to find the people who are responsible for the disappearance of so many women and children. They are not having much success, as the pterosaurs are responsible for most of those disappearances. Of course the police cannot explain this, as they are not scientists.

It is the duty of the scientists to advise the members of the public, the common people, of the existence of these prehistoric monsters, these "people eaters". I stress, that is *their duty*! The scientists are not doing *their duty*!

It may be objected that I am being too harsh with the scientists. Some people have mentioned that the scientists may not be aware of the existence of the pterosaurs, those predators that kill people in the darkness. To such kind, tender hearted souls, I can only respond: *How can the scientists not be aware of their existence?*

These animals are on the six continents, at least within the hills and mountains. In most parts of the world, they are referred to as dragons. The Chinese have named the twelve years after twelve animals, and that includes the dragon. It stands to reason that the animal is in Asia. The bible mentions various animals, and that includes the dragon. As the bible was written in Africa, that means the animal is also in Africa. As well, all scientists are well aware that this animal is a reptile, and all admit that there are *no mass extinction of reptiles!* As these flying pterosaurs are reptiles, it stands to reason that the *pterosaurs are not extinct!*

CHAPTER 3

MASS EXTINCTION
OF MEGA FAUNA

It is the scientific opinion that possibly ten thousand years ago, at the end of the last ice age, there was a "mass extinction of mega fauna", in that five species of huge animal dropped dead, because they were unable to handle climate change. These five species include the woolly mammoth, the Jefferson ground sloth, the dire wolf, the sabre toothed cat and the short faced bear. In fact, all of these animals are truly huge, or mega. What is more, they are not extinct.

For the benefit of those who are not terribly familiar with scientific terms, I will mention that "mega" means huge while "fauna" means animals. I will also mention that this theory, that of their mass extinction, is "mega" ridiculous!

As all scientists are supremely well aware, within the last hundred thousand years, we have had no less than three ice ages here in North America, all of which the "mega fauna" survived. Those same scientists are also well aware that each time the climate changed, which is to say that each time the continent cooled off and glaciers covered the land, these species survived. That merely stands to reason. It also stands to reason that each time the continent warmed up and the glaciers melted, the species also survived. The fact that these species were alive at the end of the last ice age proves this, beyond any shadow of a doubt. It also proves, also beyond any shadow of a doubt, that these species are quite capable of handling climate change. Further, they did. Those species are still very much alive.

As mentioned previously in this article, my mother in law recognized a picture of an elephant. The only difference is that the "elephant" she remembered was "hairy". That is the reason the Dene refer to this animal, the woolly mammoth, as the "hairy elephant". This is to say that the *woolly mammoth still exists!* The Dene were running from the woolly mammoth in the late twentieth century! Their one and only chance against this animal, the largest land dwelling animal in the world, was to make it to the safety of the nearest swamp!

As the swamp represents muskeg, and the animal senses that it has to avoid muskeg, the people were safe in the swamp. In winter this is not an issue, as the mammoth spends the winter in caves.

This is *not* to say that the mammoth is a flesh eater, because it is not. It is to say that the mammoth is almost certainly an intelligent animal, and as such, is very likely capable of emotion. Assuming that to be true, then it is clear that the mammoth feels emotion. The emotion it feels towards us is that of hatred. Who can blame it? Until very recently, the mammoth was wide spread across North America. It was especially at home on the prairie. Then, at the time of the European invasion, settlers appeared. The last thing a settler wants on his homestead is a five ton vegetarian. They made this quite clear by shooting every mammoth they could get in their sights.

The modern day farmers of Africa are currently in a similar situation. Their only livelihood is their crops, and the elephants take great delight in feasting and trampling the crops which the farmers have worked so hard to grow. But then those crops are generally far superior, far more nourishing, than the vegetation which grows naturally in the area. No doubt, the farmers who are suffering the loss of their crops would love nothing more than to kill every elephant which comes close to their farm. They are unable to do so due to laws. The elephants are most valuable, if only for purposes of attracting tourists.

The American homesteader of the late nineteenth century was not restricted by such laws. As the railroads pushed west from the province of Ontario, they carried settlers who cultivated the land and planted their crops. They also protected those crops, with firearms, when necessary. There was no law against killing the woolly mammoth. Those animals responded by running ever further west and north, into the relative safety of the mountains. The remnants of a once great herd, which at one time stretched across North America, is now scratching out a living in the unforgiving mountains. The problem now is to prove they exist, pass laws to protect them, and lead them out of the mountains to their former grazing grounds. Wild life photographers will line up to take their pictures.

The farmers on the prairie will not be so enthusiastic to see these mammoth. The mammoth will find themselves in a land transformed, a "land of milk and honey". They will marvel at fields of wheat, rye, corn and oats, as far as the eye can see. The woolly mammoth can see very far. They will then walk through the fences of the farmers and have a feast. The difference now is that the farmers will not be allowed to kill the woolly mammoth, as they will be protected by law, just as the elephants of Africa are protected.

The government will have to find some way to compensate the farmers for the loss of their crops, perhaps by charging a fee to tourists who delight in taking pictures of these magnificent animals. No doubt in the fall, instincts will kick in which force the mammoth to migrate south to warmer pastures, passing through the international border.

Once the animal gets the idea that it is safe, that we are not going to shoot it, it may in turn do its best to irritate us, just to "get even". It may block roads and highways, trample golf courses, even knock down fences, just to let us know that it hates us. We had best be prepared.

As the animal is so huge, locating it and proving it exists should not be difficult. It can possibly be seen from satellites, when there is no cloud cover. It can certainly be seen from small planes. The pilots who fly across these mountains, referred to as "bush pilots", no doubt see this animal on a regular basis. Equally without doubt, certain forest service personnel, those who fly across the forest districts looking for forest fires, also see the mammoth. They are careful to report the location of any and all forest fires, but equally careful to not report the location of any mammoth. Their careers depend on this.

This brings us to another species of "mega fauna", also allegedly extinct. This animal is commonly referred to as the "short faced" bear, technically referred to as "arctodus simus". The Dene refer to it as the "rubber faced" bear or the "beaver eating" bear. I refer to it as the "mega bear".

The bear is truly huge, weighing in at a ton. That means one thousand kilos or two thousand pounds. It stands one point eight meters or six feet at the shoulder. When it rears up on its hind legs it stands five meters or seventeen feet high. The claws are seven inches or fifteen to eighteen centimetres long. They use these claws to *tear apart beaver houses!*

As far as I am aware, no bear has ever been documented to tear apart a beaver house! For that matter, I have never heard of any bear that has ever tried to tear apart a beaver house, if only because they are able to sense that they cannot do this. Yet this bear makes a habit of tearing apart such beaver houses. It is that big and powerful. At least in spring, when the beaver pond is still frozen, the mega bear tears open the beaver houses. Of course the beavers dive into the water, but as they still have to breath, come to the surface in the only opening in the ice, right into the jaws of the mega bear.

To clarify, the Dene tell me that the bear has no hair on its face, so that is the reason they call it the rubber faced bear. The reason they call it the beaver eating bear is quite obvious. I have included two stories of encounters with the mega bear, by a Dene elder. As I consider the stories so important, I have decided to include them here, by permission of that elder, Seymour Isaac. These incidents happened at the time when he and his brother Francis were lads. The men were teaching them how to hunt beaver, so that the boys very likely had twenty two calibre rifles, while the men had high powered rifles, or "big guns", as they call them. I have included the story as was written by Elder Seymour. I should add, for the benefit of those who are not familiar with imperial units, that "six or seven inches" is the equivalent of fifteen or eighteen centimetres. Also, seventeen feet is the equivalent of approximately five metres.

But now let the elder tell his story, in his words:

The title is Beaver Eating Bears

I remember the first beaver eating bear that was shot and killed way back in 1953 or 1954. Grandpa Keom Pierre and his stepson Larry Pierre and William Poole and Jimmy Dennis, Charlie and I, who were all young boys at the time, not yet in our teens, were trapping beaver. We had been doing so for five days and headed up to Shovel Creek Pass on the far side of the trap line which Grandpa Keom called 15 Mile Cabin. Around the first week of April, there was still four to five feet of snow, so we used snowshoes. We had set beaver traps under the ice and found some places that had exposed spring water.

We knew that Davis Creek flowed past 15 Mile Cabin only about three miles away. Keom and Larry also knew that the Davis always opened up early, so the boys were told to bed early. At first light, we would head to Davis Creek while the snow crust was still frozen. The next morning, we all got ready to travel on top of the crusty snow and took three of Grandpa's hunting dogs with us. The dogs were named Tez, Pup and Silver. We travelled on and finally came to the creek. The dogs started barking a commotion like they had seen something they did not like -and they certainly did. In front of us stood the biggest bear we had ever seen; it was bigger than the biggest bull moose we had ever seen. Then the fun began.

They started shooting and Keom just shouted, "Don't hit the dogs," but the dogs had the advantage because the bigger bear was breaking through the four inch snow crust and the dogs were on top of the snow and not sinking through. Since the bear was breaking through the crust, it could not move fast. Once that bear saw us, the bears concentration was on the dogs. The bear tried to get them, and the men tried to move away from the bear as they were shooting. After twelve or thirteen well place shots, the bear finally dropped. Grandpa Keom had used an 8-millimetre war gun, Uncle Larry had used a 303 British war gun, and Jimmy fired his brand new 30-30 Marlin lever action. But it was William, with a single shot .22, who took the bear down that day.

While the bear was being shot at, it thrashed everything and anything that was in its way. It ripped out little trees; willows flew left and right. More than a few times, the bear stood up and was a massive seventeen feet tall. Man, did he stink! After the chaos was over, the bear was skinned. Its claws were six to seven inches long. The hide could not be saved because the side that he slept on was ruined from his shoulder to his hips from rubbing.

The second story from this Elder is titled "Beaver Eating Bears in Akie".

In 1958, my brother Francis Isaac and I were up the Akie River to do some spring beaver trapping with Uncle Angus Pierre and Uncle Mac Pierre and the whole family. We just loved Grandma Elizabeth and the great stories that she always told.

At about mid-May, the five of us being myself, Uncle Angus, Auntie Lucy, Uncle Mac and my brother were going up the Akie. Old Grandma was left to babysit at the time.

We started to travel up the river. As I recall, the Akie is a very difficult river to shoot beaver on in the spring. At that time of the year, the river is so low it almost becomes a creek. When the river is low, the beavers do not come out until dark, so we mostly used traps. We did pretty good in trapping on a daily basis and trapped three or four beaver each day. When we baited the traps, we used poplar tops mixed with beaver castor and used a little of our own oil mix, which worked very well in the traps as bait. Since we camped along the way, we would set our traps and enjoy ourselves until it was time to check the traps.

As we cut trail along the way, Uncle Mac told us to look out for ourselves as there was a lot of bears and grizzlies in the area, especially where we were cutting trail. During our way along the trails, we encountered bears and scared them away. We didn't encounter any problems until that one fine day on the way back. We passed the second trapping cabin and went right on up to the area that's called the big bend in the Akie River.

Our Sekani ancestors had a name for that place and called it Big Bag Yorks. Uncle Mac shared some history of that place. Our people used to stop there to dry meat, and the women picked berries and medicinal plants. Big Bag Yorks was a nice place to camp. It got its name by being a place where most of the valleys could be seen. Creeks ran into the

Akie, and there was plenty of game in that area. This place should be marked as a historical marker on our map as a secret place.

We had camped in the area for about three days before heading down the Akie. On our third day of camping, we encountered what our ancestors called the beaver hunter.

At the time, Uncle Mac was setting a beaver trap. Uncle Angus told him that we would travel on and wait down at the point for him. We came to a place with a lone sandbar right by the bluff. The water must have been quite deep as I remember an eddy in the green, pristine water.

In that area, we saw a beaver jump in. Uncle Angus took out his gun and went down the bluff to look for that beaver. Right across the river, we saw a dry slough that ran right into the Akie. I thought we had seen a moose and Francis thought that too. All of a sudden, Francis looked at Uncle Angus and wanted to know who was on the sandbar? At first, we thought maybe it was someone, but it was a beaver hunting bear coming towards us on the sandbar. It was only fifty feet away from us. Uncle Angus was ready, and grabbed his gun and shot. Those first two shots were so fast and accurate that it sounded like one shot! Uncle Angus shot four more shots then ran and knelt by a tree. He started shooting more. Although Uncle Angus had a single shot 30-30, he was shooting out shell like an automatic. Oh, the action! In the mean time, when all this shooting was going on, we could hear that beaver hunting bear yelling and hollering, crawling, rolling around, and scooping up paws full of gravel. The huge bear stood up. After fifteen shots, it had to have felt the shots and started walking on the little island. One 30-30 and one 303 British had hit him. Finally, we heard his last, big growl. Some of us wanted to check the bear out, but Uncle Angus noted it was getting late, and we knew it was dead.

The next day, we got up early and headed back home. We told old Grandma and the kids about the encounter with the beaver eating bear.

She told us that she knew about it. As long as everyone was okay, she was happy.

Between the four of us, in twelve days we got forty five beaver and twelve muskrats!

After we were back, Billy and Art Van Somers came up and brought supplies. They brought a parcel and a letter for me from Dad saying he wanted both of us back to town. He also mentioned he had missed us very much and that he was staying with our sister Louise and her husband Wilson in Summit Lake. When they came back, we went to town with them and shared our proud hunting story.

To continue with our list of mega fauna, the Jefferson ground sloth is named after Thomas Jefferson, the third president of the United States. The Dene refer to this animal as the giant beaver.

In all fairness to Thomas Jefferson, the man was a genius, truly talented. As well as being a diplomat and a politician, he was a scholar, inventor, scientist and architect. He personally designed the building on his plantation of Monticello. It is considered to be an absolute masterpiece. This was the man who first described the animal which is named in his honour, the Jefferson ground sloth.

This in no way changes the fact that the man was a slave owner, a true psychopath. Nor does it change the fact that the scientists are supremely well aware that the animal was alive two hundred years ago, as it is today. Yet to this day the scientists insist that the animal is extinct!

Another example of the hypocrisy of the scientists is the supposed extinction of the dire wolf. The wolf is huge, the size of a deer. It may weigh perhaps one hundred fifty pounds or seventy kilos. The Dene refer to this animal as the "wilderness wolf". It is well named, indeed "dire", as this animal is not afraid of us! They come into the village after sundown and prey upon dogs. Those dogs which are tied up are particularly easy prey animals.

Even the scientists admit that the dire wolf still exists, even though they claim that it is extinct. As they phrase it, the dire wolf exists, but only as "a remnant population of an extinct species". Such nonsense is now referred to as "alternative facts", although I have another name for such foolishness. I call it what it is. A pack of lies!

This brings us to the sabre toothed cat, technically referred to as smilodon fatalis. This huge cat, the size of a siberian tiger, loves the grass land, so is not located in the mountains. It loves the prairie, and in fact has been seen on numerous occasions within the city limits of Milwaukee. For that reason, it is commonly referred to as the Milwaukee lion. There are numerous pictures and videos of the animal, posted on the internet. It is a female, which explains the reason the animal lacks huge teeth. Such teeth are characteristic of the males of the species only. The males have evolved those "sabre teeth" for purposes of display.

CHAPTER 4

OGOPOGO AND SWIMMING REPTILES

The same scientists who swear that there was a mass extinction of dinosaurs, and a mass extinction of mega fauna, as well as a mass extinction of flying reptiles, pterosaurs, also maintain that there was a mass extinction of four orders of swimming reptiles. In particular, it is their belief that the orders of mausosaurs, long necked plesiosaurs, short necked plesiosaurs and ichthyosaurs all dropped dead, for reasons which no one can imagine. This is all the more remarkable, as those same scientists also maintain that *there are no mass extinctions of reptiles!*

They have chosen to ignore various eye witness accounts, many of which are highly reliable, of "lake monsters" which are seen in certain bodies of water, especially in Loch Ness, Britain, and Okanagan Lake in North America. For some reason, the animal is referred to as "ogopogo".

The accounts of these sightings are somewhat scattered and vague, if only because most people are reluctant to go public and risk being branded a nut. Then too, there is no shortage of people who delight in spreading confusion. As they have nothing to gain by these acts of idiocy, such actions defy comprehension. Possibly it is just their nature to cause trouble.

That being said, as best I can gather from highly reliable sources, which is to say the Indigenous Elders, this animal is very much alive and supremely dangerous. It is also nocturnal, spending the day light hours in under water caves. It rarely comes to the surface during the day light hours. Then again, many people have reported seeing it in Okanagan Lake, so it definitely comes to the surface, if only on occasion. Possibly the reason it is seen so frequently in Okanagan Lake is because so many people live around the lake.

I suspect the reason these animals avoid the light is the same reason the flying reptiles avoid the light. They too may be facing competition from raptors, birds of prey. It is quite likely that each time these animals come to the surface, the raptors land on their backs and nibble. Assuming the hides of these animals are tender, that would explain their reluctance to surface during the day. The wound does not have to be terribly deep and severe to prevent infection. A relatively minor wound could conceivably be fatal. That is a good reason to avoid the light.

As animals which are nocturnal, they face the same problem the pterosaurs face, that of a matter of displaying during mating season. It is very likely they have come up with the same solution, that of glowing in the dark. That would explain the reports of underwater lights, in various bodies of water, fresh water lakes as well as oceans. That makes more sense than suggesting these are the underwater docking sites for spacecraft!

I have long maintained that the best way to locate them, or at least the safest way, is to look for their nesting site, the beach where they lay their eggs. As reptiles, they reproduce by laying eggs, and of course eggs need heat in order to hatch. Just as turtles lay their eggs in the sand of the beach, so too these reptiles lay their eggs in a similar manner. Or so I thought.

It came as a rather rude shock to be told that there is no naturally occurring sand on the beaches of Okanagan Lake. The only sand on the beach has been trucked in, to be used by tourists. That was not at all what I wanted to hear.

Yet all reptiles reproduce by laying eggs, so how does the reptile in Okanagan Lake keep the eggs warm? Unless it is not a reptile at all, but a whale called basilosaurus cetoides, a mammal long thought to be extinct. That is one possibility I cannot rule out.

On the other hand, a little more research revealed that one order of swimming reptile, ichthyosaurs, are thought to be warm blooded animals which reproduce by giving birth to live young, head first. *So how can they be classified as reptiles?*

Yet they are classified as reptiles, even though reptiles are *cold blooded animals which lay eggs!* More to the point, how is it that I was surprised by this? You would think that by now I would come to expect the scientists to contradict themselves! Me and my starry eyed optimism!

Almost all eye witness accounts agree that this animal resembles a huge snake, but swims like a mammal, not a reptile. This sounds ever more plausible, if only because it is truly a mammal, which is classified as a reptile. As I am not terribly interested in the classification of animals, it is a subject of indifference to me. I am very much concerned with proving the animal exists, mammal or reptile.

Be that as it may, the fact that the ichthyosaur gives birth to live young head first, implies that the animal has to come out of the water to give birth. Otherwise the youngster will drown in the process of being born. Dolphins give birth in the water, but tail first, and as soon as the head emerges, the newborn immediately scoots to the surface for a breath of fresh air.

In the case of the ichthyosaurs, they are forced to give birth on land. Almost certainly, they use the same bit of land to give birth each year. Now the problem is one of locating that birthing ground. Without doubt, the best, most reliable source of information is the local Indigenous Elders. They can very likely point out the birthing ground of the animal.

Once that birthing ground is located, then it a simple matter of placing a camera on the beach. As well, the "after birth" will provide a great deal of DNA. The problem is one of grabbing that DNA before the predators have a chance to consume it. Then again, if the animal gives birth in under water caves, then the problem of proving the existence of the animal just became much more difficult.

Assuming the best case scenario, and the animal in Okanagan Lake is proven to be the ichthyosaur, that still leaves three orders of swimming reptile, thought to be extinct.

For the benefit of readers who are not scientists, I will mention that within each order of animals there are possibly dozens of species. For example, in the world today we have twenty three species of crocodiles, while we have possibly one hundred species of turtle. The difference is that no species of crocodile can handle any cold whatsoever, while many species of turtle have evolved that ability. No doubt there are other reasons to help explain the wide spread existence of turtles.

Concerning the swimming reptiles, or at least the other three orders of swimming reptiles, no doubt some species have evolved the ability to live in salt water, while other species have evolved the ability to live in fresh water. It is very likely they all have tender hides, and are nocturnal, as other wise they would be seen more frequently. I can think of no other reason for not seeing them on a regular basis.

It is possible that "ogopogo" of North America and the Loch Ness monster of Britain are the same animal, or possibly not. The animal in Okanagan Lake may be unusual because there is no sand along the lake shore. As yet we have no way of knowing. We do know that there are a great many lakes in North America which are reported to contain "lake monsters". Now it is up to members of the public, common people, to become active and join the search for these reptiles.

Each of these reptiles is an apex predator, so it is not likely that any particular lake will contain more than one species of such predator. Such predators take great delight in killing members of another species. They do not tolerate competition. It is the key to survival. The first step in locating these animals, in any particular lake, is to first determine if there is any sandy beaches. If so, check with the local Indigenous Elders, with a view to determining any possible beach where the animals lay their eggs. Or the animal may be an ichthyosaur, and prefers to give birth on the beach.

If the animal is a reptile and lays eggs on the beach, then just before the eggs hatch, raptors will gather. They will be seen sitting on the branches of trees, waiting for a hatch and subsequent feast. As well as the Indigenous people, those who work on the lake, as well as people who live along the beach, may have seen the gathering of raptors, on a particular beach, without giving it much thought. That may be key to locating the nesting site of the swimming reptiles. Once the nesting site is identified, then it is a simple matter of keeping an eye on the beach, waiting for a gathering of raptors. As soon as the raptors appear, set up a camera and join them. As soon as the hatch begins, record this. Grab a youngster or two, in the interests of science, in order to prove, beyond any shadow of a doubt, that the animals exist. Granted, we should not interfere with the hatching of any animals, but in this case, it is correct to make an exception.

Equally without doubt, many species of swimming reptiles have adapted to life in the ocean, salt water reptiles. Locating the nesting sites of these reptiles is far more difficult, if only because they have far more beaches available to lay their eggs. On the other hand, such beaches are a popular place to build houses, so that there are far more people who can be counted upon to spot the gathering of raptors. Now that most people have computers and there is no shortage of young people who know how to use those computers, we can only suggest that those people, young or old, put those skills to good use.

There are four orders of swimming reptile just waiting to be discovered. Now is the chance to take part in a major scientific break through.

CHAPTER 5

SASQUATCH OR BIGFOOT

All scientists are agreed that of all the species of human that have ever evolved, we are the one and only species of human still living. In this too, they are mistaken. Here too, all reports, many of which are highly reliable, of giant naked hairy humans, are carefully ignored. As well, countless plaster casts of footprints are dismissed as fakes.

I maintain that there is a separate species of human currently walking the earth. Further, they live among us, here in North America. That species is technically referred to as gigantopithecus, although they are most commonly referred to as sasquatch or bigfoot. The Dene refer to them as "stink people", a title which is less than flattering. I refer to them as Giants, because they are truly huge.

The scientists believe that this species of ape evolved many years ago in Asia. As it stood ten feet tall or three meters and weighed twelve hundred pounds or five hundred kilos, it was by far the largest ape ever to walk the earth. In fact it was twice as big as a gorilla. But then, the scientists maintain that around one hundred thousand years ago, this huge ape, gigantopithecus, dropped dead, for reasons which no one can imagine.

The scientists are mistaken. Gigantopithecus did not go extinct. They first became bipedal, which means they started to walk on two legs, and then they developed the opposable thumb. This is to say that they can touch their fingertips with their thumbs. In other words, they evolved into humans. It is my contention that apes which first evolve bipedal locomotion and then evolve the opposable thumb, are nothing other than a species of human. As humans, they then began to bury their dead. That is the reason no recent remains of these apes has been discovered.

It should come as no great surprise to anyone to find out that a great many scientists disagree with me.

As members of a separate species of human, they deserve our utmost respect. They have every right to live their lives as they see fit. Our laws do not apply to them. We have no right to interfere in their way of life. We certainly have no right to hunt them, as so many people are hunting them, as if they were vermin, determined to kill them. Such a killing can be seen as nothing less than an act of murder, completely unjustified. The fact that many of the people who are hunting them are doing this with good intentions, does not change that fact. Those people think that this is the only way to prove they exist. The end does not justify the means.

I maintain that these people, the Giants, are members of a hunting-gathering society, so that they are constantly travelling. To stay for any considerable length of time in one area would quickly result in the exhaustion of the food supply, if nothing else. For that reason, they are constantly moving, following the herds. As well, the wild plants are harvested in the proper season, in the proper location.

In the spring, they travel north, in order to take advantage of the abundant supply of eggs of birds, as well as the birth of numerous animals, such as deer, moose, elk and caribou. They also harvest the wild berries as they become ripe. In the fall, they travel south, in order to take advantage of the warmer temperatures, as well as different food supplies. The further south they travel, the more settlements they pass through. The people who live in those communities take note of the tracks they see in the morning. They know full well that something big had placed those tracks the previous night, while overlooking the fact that at the time the tracks were being placed, the dogs were not barking. This silence of dogs is significant, as dogs are known to bark at all animals known to science. As the dogs are terrified to the point of silence, in the presence of Giants, this should give all of us some idea of the size, strength and sheer terror their presence inspires in dogs. If nothing else, this is reason for giving them our utmost respect.

As many people are hunting them, their lives are being made quite miserable. They are constantly on the run. This constant harassment can result in something more than inconvenience. They are members of a hunting-gathering society, so their food supply is completely unstable. It varies according to the season and changes in the weather. They have no emergency food supply to fall back upon. They live constantly on the edge of starvation. Hunger is a constant companion.

In addition, with our constant expansion and industrial development, we are destroying their hunting and gathering grounds. This industrial development includes dam building, logging, mining and road building, which makes it difficult for the young males to travel to different bands of people, to whom they are not related, in order to find wives. This is the one and only way to avoid in breeding, and is the only way to avoid the extinction of the species.

This is not an exaggeration. The current European nobility can testify to the fact that inbreeding can lead to disaster. All of them are related, and their off spring, the heirs to the throne, can at best charitably be referred to as "simple souls". Anyone who doubts this has only to face the fact that the queen of England and her husband are closely related. Ever since the birth of their first child, they have no doubt regretted the fact that they "tied the knot". The heir to the throne of England is a constant source of embarrassment to the queen. For that reason, it is now fashionable for the nobility to insist that their young people marry "commoners". This is not to say that they find us any less contemptible, as they do not. They have merely been forced into this.

The "royal watchers" are now speculating that the queen has already decided that the next king of England will not be her first born male child, but her first born male grand child. That child is not as deeply inbred as his father.

The point of this is to drive home the fact that inbreeding is certain to lead to disaster. We have no way of knowing the population size of the Giants, but we do know that as members of a hunting-gathering society, their numbers are limited by their food supply. This is another way of saying that there cannot be a great many of them. If their population is reduced to a very low level, or if the males are unable to travel to other groups in order to find wives, then the species is certain to go extinct.

To be responsible for the extinction of another species of human goes beyond genocide. Yet unless we change our ways, we could be guilty of precisely that.

For that reason, it is urgent that we prove the existence of these Giants. Many people have been trying to do this, but they are going about it in the wrong way. A different approach is required. Instead of chasing them, attract them. They have no reason to trust us, as we are quite fond of shooting them. It will take a while to convince them that we mean them no harm.

The only members of our species they trust is the Indigenous people. The Giants and the Indigenous people respect each other. They leave each other strictly alone. Live and let live. On the reserves of Vancouver Island, they share the beaches. The Indigenous people enjoy the beach during the day, while the Giants enjoy the beach during the night. It is on the beaches of the reserves that we can meet them.

The key to this historic meeting lies with the Indigenous Elders. They have about as much reason to trust us as do the Giants. I can only suggest calling a Reserve on Vancouver Island and requesting a meeting of the Elders Society. Assuming they agree, then arrange a feast for the Elders. Be sure to bring in food those people love. As well, buy gifts, such as tobacco, Hudson Bay blankets, electric blankets and tanned hides. The management can advise the gifts the people appreciate. Spare no expense, as we are looking at a major scientific break through. Make sure that food and gifts are taken to the Elders who are too feeble to attend the meeting. Respect is essential to the success of this project. With that in mind, offer to hire a translator. It is very likely that all the Elders understand English, but a little show of respect goes a long way.

After the Elders have eaten and received their gifts, explain to them that we are interested only in meeting the Giants. We mean them no harm. On the contrary, there are countless people who are interested in killing them. The one and only way of preventing this is to have laws passed to protect them. Further, the only way this is about to happen is by first proving they exist. That is where the Elders come in. It cannot be done without the help of the Elders.

With that in mind, mention that the ideal place to make contact with the Giants is on the beaches of the Reserves. This can only happen with the blessings of the Elders and the cooperation of the young Indigenous people. The Elders know the time the Giants are on the beach, and with the help of the young people, gifts can be placed on the beach, close to sundown, for the Giants.

As the Giants are human, there can be no doubt that they love the same things that we love. That includes meat, vegetables and fruit, raw as well as cooked. It also includes cosmetics and mirrors, preferably small metallic mirrors. Decorations of metallic ornaments, tied together with dental floss, is also an idea, as is burlap bags.

As for those who complain that these are merely "beads and trinkets", I can only respond that you are absolutely correct. May I suggest that our goal is to meet these people, to prove they exist. We want to impress upon them that our intentions are honest, and gifts of platinum, gold and silver are not about to do a world of good. They have no use for such items. No doubt all of them want to know what they look like, and most members of our species wear cosmetics and decorations of one sort or another. As the Giants are just as human as we are, it is reasonable to assume that they love the same things we love.

Perhaps it would be best if government officials are not involved in this science project. Such officials tend to have their own agenda, which is frequently at odds with the goal of the project. At best, such officials merely spread confusion. At worst, they sabotage the operation. It is a mistake to expect anything better from such people. Then again, it would be so nice if even one of them was to surprise me by doing something intelligent.

It is only after we prove the existence of these people, the Giants, that we can focus on getting laws passed to protect them. This can only be done through the United Nations, as the Giants have no knowledge or concept of international borders. Besides, the local authorities have other priorities.

CHAPTER 6

CAPITALISM AND OUR CIVILIZATION IN DECLINE

It is remarkable to think that for over thirty years, I have been working on this, my little science project, as I refer to it. During that time I have been challenging various scientific theories, mainly searching for huge prehistoric animals, which I am convinced are not extinct. Perhaps my greatest surprise was to find that so many scientific theories can best be described as glorified fairy tales. It is sad to think that the state of science has been reduced to such a pitiful state.

Now there are a great many people who look back upon the world before the time of Newton as "the good old days". It was certainly a much simpler time. It was accepted, and in fact it was insisted, that the world was the centre of the universe. The answers to all questions were to be found in the bible. Anyone who challenged this belief could be charged with heresy and burned at the stake. It was this threat of execution that served to limit scientific advancement.

All of that changed with Newton and his three laws of motion. We now know that the earth is anything but the centre of the universe.

Other scientific break throughs followed. In particular, Darwin did for biology that which Newton did for physics. Not everyone was too happy about this. To this day, there are a great many people who resist change. In particular, those who are referred to as "fundamentalists" or "creationists" deeply disapprove of the theory of evolution. The beliefs of these people, common people, members of the public, must be respected. It is their religious belief that people were created and the theory that people evolved from apes is sheer nonsense. I stress that it is up to everyone to respect that belief.

That being said, it is important to separate science and religion. Religious beliefs have no place in the class room. To teach "creation science" in the class room, as an alternative to the theory of evolution, cannot be tolerated. After all, "creation" is a belief and not a science.

It is important that we show the utmost respect to those who have very strong religious beliefs, and distinguish those people from the reactionaries. Those are the people who want everything to stay exactly the way it is.

As for those who think that such reactionaries are rather quaint, harmless sorts, feel free to think again. Bear in mind the fact that over the last several thousand years, numerous civilizations have come into existence, flourished, risen to a peak, and then fallen into decline. This decline in civilizations is not an act of God. It is an act of people. Honest, hard-working people build. Then along come the reactionaries, those who are lying, thieving people, and they do their best to destroy.

Our ancestors have worked all their lives to build a better world, for the benefit of their descendants. Many of them made the ultimate sacrifice, while others worked themselves into an early grave. Without doubt, they built this civilization, and equally without doubt, it has now passed its peak and is in decline. I repeat, this decline is an act of people, reactionaries, not an act of God. Unless we take action now, all the work our ancestors have done will be for nothing. It is up to us to honour their memory by following in their footsteps.

As for those who are skeptical, may I suggest as an example, the ancient city state of Sparta. At one point, the Spartans were superb warriors, feared by all their enemies. At their peak, they could field an army of ten thousand soldiers. Then the rot set in, and the city state of Sparta fell into decline. Within a hundred years, they were barely able to field an army of one thousand soldiers. Sparta was merely a shadow of its former self.

After all these years, the precise details of the decline of Sparta are blurred by the mist of time, so to speak. Still, scholars are convinced that Sparta was ruled by a council of elders, men over the age of sixty. Each council ruled for one year, and during that time, the council could do anything it pleased. Their word was law. But at the end of that one year term, each and every member of the council went on trial for any and all crimes they may have committed, during the previous year. They were presumed guilty until proven otherwise. If convicted, such offenders were immediately executed. There was no court of appeal. Spartan justice was immediate and terminal.

The elders who were elected to the council could avoid trial and execution only by doing nothing, while serving on the council. This they did supremely well. For many years, the city state of Sparta remained frozen in time. The council of Sparta made sure nothing changed within Sparta, while the world around Sparta changed dramatically.

The warriors of Sparta loved hand to hand combat, but only hand to hand combat. They had no use for projectiles, whether from archers, javelin throwers or soldiers with slings. They considered such soldiers cowards, and were not about to stoop to such depths. Among other things, the ruling council of Sparta made sure that their warriors engaged only in hand to hand combat.

The enemies of the Spartans had no objection to this. In fact, they welcomed this approach, by their enemies, the Spartans. Their enemies also welcomed improvements in metal working, various weapons and tactics. The Spartans, by contrast, were stuck in the old way. Their civilization declined very quickly.

The decline of the Spartan civilization was by no means an exceptional event. It merely provides a very clear cut example of a civilization which rejects change.

A more recent example is supplied by the Chinese. Several hundred years ago, the Chinese civilization was far more advanced than any other civilization in the world. Among other things, the Chinese invented gun powder and steel, crossbows and automatic weapons. Gunpowder alone is a decisive weapon in civilized warfare. Then the Chinese civilization fell into decline.

The point I am trying to make is that our civilization is also in decline. It has risen to a peak, and now the people who oppose all change, the reactionaries, are in charge. They are doing their level best to destroy all that which our ancestors have created.

This retrograde trend in science is nothing more than the result of the monopoly capitalists, imperialists, all of whom are reactionary, interfering in science. They refer to themselves as "entrepreneurs", rather than capitalists, as if changing the name changes the nature of the beast! They live in the past, and are determined that nothing will change. The science books have been written and are not to be rewritten! The same is true of the history books and all text books used in schools! The fact that both science and history books are filled with distortions and outright lies does not impress them. The current up roar in America, and not just America, concerning the glorification of slave owners serves as a reminder of the mind set of reactionaries.

This is precisely the very course of action which has proven so successful in the past, in terms of bringing down civilizations. Now the modern day reactionaries, in the form of the capitalists, are determined to destroy our civilization. They can and must be stopped.

The capitalists have university training set up so that tuition costs a great deal. In order to earn a science degree and work in any particular field of science, most students are forced to go deeply into debt. Those who earn such science degrees are typically saddled with huge student loans. Then it is a matter of getting a job working in their chosen field of science and at least meeting the payments on the student loan. Frequently, the loan is so huge, there can be no question of paying for it entirely. The borrowers are merely expected to meet the interest payments.

These huge student loans give the capitalists great leverage over the scientists. Any and all scientists who wish to pursue a career in science must not "rock the boat". The scientific theories, as presented in the text books, are not to be challenged! The penalty for "rocking the boat" is commonly referred to as "career suicide". Such offenders are "black balled", not allowed to earn a living in any field of science, while still being saddled with a huge student loan. As a result of this, scientists are forced to embrace theories which they know to be false, as that is the only way to survive. That is nothing less than degrading.

The exception is the few students who are wealthy enough, or at least are members of families which are wealthy enough, to pay for the school tuition. Such people are supremely class conscious, usually members of the petty bourgeois, middle class, and know better than to "rock the boat". They are supremely well aware that the bourgeoisie are in charge, and such people do not tolerate any challenge to their authority. Remaining silent concerning the existence of these huge animals, among other things, is the price of success.

As a result of the silence of the scientists, most common people, members of the public, are blissfully unaware of the existence of these prehistoric monsters, including the pterosaurs. It is the pterosaurs, in particular, that frequently prey upon people. They hunt in open areas, in the darkness, and as predators, have a keen sense of smell. It is characteristic of predators that they are far more likely to attack when they smell blood. The pterosaur is no exception. That is the reason that it mainly preys upon girls of child bearing age. It also preys upon children, as children are easier to pick up and carry away.

No doubt the scientists are aware of the existence of these huge species. Equally without doubt, they remain silent in order to protect their careers. I have no career to protect, as I committed career suicide many years ago. It was the price I paid for "rocking the boat". Those days are long gone, so the capitalists cannot threaten me with career suicide. Besides, my days of "rocking the boat" are over. Now I am blowing the boat right out of the water! Soon a great many careers will be ruined, or at least severely compromised, as I prove the existence of these animals.

As the scientists either cannot or will not challenge any scientific theories, it is now up to the members of the public, common people, to "blaze the trail". To prove the existence of these species is not difficult. The scientists are careful to avoid locating them, even by accident.

Then again, the revolutionary motion which is sweeping the world is growing ever stronger. It is just a matter of time before the scientists are affected by the revolution. When that happens, we can expect them to rise up, band together and demand change.

No doubt many readers are of the opinion that I am overstating the case. So for the benefit of such simple, honest, misguided souls, may I refer you to an American film maker, Michael Moore. Mr. Moore is a most patriotic American. He has produced a number of documentaries, of which I consider Capitalism: A Love Story, to be his masterpiece.

Michail Moore is certainly not a Marxist, so he refers to the monopoly capitalists, the imperialists, the bourgeoisie, as the "one percent", 1 %. That is an expression the common people have coined, as a means of implying that those people, the "super rich", are members of a very small minority, one in a hundred. In fact, their numbers are far less than one in a hundred, but that awareness, of the working class, the proletariat, of themselves as separate from the "super rich", is a step in the right direction. It is a step towards class consciousness. They are drawing a line, so to speak, if only instinctively, between them, the working people, the proletariat, and the monopoly capitalists, the bourgeoisie.

With that in mind, I have tried to summarize the facts presented on the documentary. Bear in mind that Moore makes no reference to class content.

In his film, Moore managed to get his hands on a secret Citibank memo concerning the plan of the "one percent" to rule the world. In fact, Citigroup wrote no less than three memos to their wealthiest investors. The bank has concluded that the United States is no longer a democracy, but a plutocracy- a society controlled exclusively by and for the top one percent of the population. As they point out, the top one percent of the citizens have control of as much wealth as the bottom 95 percent combined, and the gap is growing. Their biggest concern is that the 99 percent may demand a more equitable share of the wealth. The 1 percent are now the new aristocracy, but their big concern is that the 99 percent may revolt. They lament that the 99 percent still have the right to vote.

Moore also makes the point that the constitution makes no reference to capitalism or to the right of the capitalists to make a profit. He could further have made the point that the Founding Fathers of the United States, those who wrote the Declaration of Independence, went much further. As they stated it: "We hold these truths to be self evident, that all men are created equal, that they are endowed with their Creator with certain inalienable rights...whenever any form of government becomes destructive of those ends, it is the right of the people to alter or abolish it, and to institute a new government".

The Declaration of Independence gives the Americans the right to *abolish any government* which does not represent them! This is a point which must be stressed to all Americans! They are one of the few people, if not the only people, who have that right! But then they have a revolutionary history of which they can be most proud. Now it is a matter of building upon that revolutionary history.

No doubt the capitalists, the bourgeoisie, would love nothing better than to scrap the constitution and the Declaration of Independence. Those revolutionary documents stand for democracy, majority rule, and the bourgeoisie is dead set opposed to such democracy.

As for those of us who have lived all our lives under capitalism, which is almost everyone, it is only natural to consider this as the normal state of affairs. Yet it is any thing but the normal state of affairs. In fact, it is supremely abnormal.

Capitalism first came into existence roughly three hundred years ago. It was the industrial revolution that gave birth to capitalism. Before that time, there were no capitalists, bourgeoisie, and of course there were no workers, proletarians. The point being that capitalism is a relatively new creation. Further, it was created by people. Capitalism was not an "act of God". It was created by people and it will be destroyed by people. Capitalism will be destroyed by one of the classes of people it created, the working class, the proletariat.

Marx and Engels documented the dual nature of capitalism in their landmark work, The Communist Manifesto. They reveal to all the progressive aspects of capitalism, at least in its early stages, as well as its reactionary features. It is to be hoped that all readers will read that essential work of scientific socialism, the only true socialism.

The Communist Manifesto was written in 1848. Of particular interest is the introduction to that work, by Engels, in 1883, shortly after the death of Marx. As Engels stated it:

> "The basic thought running through the Manifesto -that economic production, and the structure of society at every historical epoch necessarily arising therefrom, constitute the foundation for the political and intellectual history of that epoch; that consequently (ever since the dissolution of the primeval communal ownership of land) all history has been a history of class struggles, of struggles between exploited and exploiting, between dominated and dominating classes at various stages of social evolution; that this struggle, however, has now reached a stage where the exploited and oppressed class (the proletariat) can no longer emancipate itself from the class which exploits and oppresses it (the bourgeoisie) without at the same time forever freeing the whole of society from exploitation, oppression, class struggles -this basic thought belongs solely and exclusively to Marx."

Within a few years, at the turn of the twentieth century, capitalism reached the stage of monopoly, otherwise known as imperialism. At the monopoly stage, capitalism assumes features which are different from those of competitive capitalism.

It was Lenin who examined capitalism in its most advanced stage, that of monopoly capitalism, in his ground breaking work, Imperialism, the Highest Stage of Capitalism. He made it clear that imperialism, which is monopoly capitalism, is completely reactionary. As we are at the stage of monopoly capitalism, we can expect nothing progressive from the monopoly capitalists, the imperialists. On the contrary, we can expect nothing other than the decay of our civilization. That work too is required reading for anyone who wants to prepare for the approaching revolution and the subsequent Dictatorship Of the Proletariat.

Marx was absolutely right when he made it clear that "economic production, and the structure of every society at every historical epoch necessarily arising therefrom, constitution the foundation for the political and intellectual history of that epoch".

It should be stressed that our "economic production" is now socialized, so that our political structure must follow suit. Now it is a matter of establishing socialism, but on a scientific basis. In other words, socialism must be Marxist, as it has been proven that utopian socialism does not work. The monopoly capitalists, the bourgeoisie, the imperialists, are not about to surrender their hard stolen wealth and power without a fight. That is the reason they must be overthrown and then crushed under the Dictatorship Of the Proletariat.

I consider the quest for these huge species to be part of the revolutionary movement. We are entitled to our wildlife. It is part of our heritage. Just as people of Africa have the right to elephants, so too do we have the right to our woolly mammoths.

It is frequently stated that any great scientific break through will have to be made by youngsters. Those who hold that belief use the example of Newton. They point to the fact that he did all of his scientific work before the age of twenty one. That is certainly true, but it does not mean that the rest of us must retire our brains at that age! I am certainly not a youngster, and am physically not capable of doing that which I did many years ago. But then my brain has not atrophied. I just hope that others, young and not so young, will be motivated to take part in a scientific breakthrough. It is not something you will ever regret.

It may help to consider this quest for these magnificent species as being part of the revolutionary movement, as that is precisely what it is. The bourgeoisie is dead set against this, if for no other reason than that it represents change. As reactionaries, they are dead set against change.

CHAPTER 7

BASILOSAURUS: BOTH THE LOCH NESS MONSTER AND OGOPOGO

As is well known, the scientists maintain that the "dinosaurs" went extinct sixty five million years ago. They also maintain that the "great sea lizards and the snake necked plesiosaurs were also dying out", at that time. It is no secret that I consider this to be nothing other than a scientific fairy tale. I have documented that most of the animals which have been classified as dinosaurs were nothing other than birds, while others, such as the flying pterosaurs, were reptiles. I also maintain that the "great sea lizards and snake necked plesiosaurs" are still alive, as are the pterosaurs.

With that in mind, I suspected that the animal in Okanagan Lake, the so called "Ogopogo", was nothing other than one of those swimming reptiles. Yet I was mistaken, as I have recently determined that the animal in Okanagan Lake, is nothing other than Basilosaurus, a huge whale, a whale with legs, a walking whale.

As I have documented in a previous article, this makes the task of proving the existence of this animal much easier. Basilosaurus is a predator, but not a carnivore. It is an omnivore, which is to say that it consumes flesh and vegetation. It is also nocturnal, in that it spends most of the daylight hours in caves, which are accessible from the water.

I should add that common people refer to these caves as "underwater caves". They are not under water, but it is perfectly acceptable for common people to refer to those caves in that manner. It is up to us, professional people, to determine the fact that those caves are above the water line, so that the caves provide the animal with shelter, warmth and a place to rest.

Then after sun down, the animal comes out of the water and consumes vegetation, mainly grass, no doubt. That makes the task of proving the existence of the animal simplicity itself. It is simply a matter of setting up trail cameras, on the edge of the meadows, adjacent to the lake. In the moon light, we should be able to get some fine pictures of the animal. As it is the largest of all land dwelling animals, it should not be difficult to locate.

The implication is that I was also mistaken when I suggested that the woolly mammoth is the largest of all land dwelling animals. Such is life. Live and learn.

My mistake was instructive so that possibly others can learn from this. Common sense told me that a fresh water lake could not possibly support a population of huge predatory swimming mammals. I still maintain that I was correct. So I assumed that the predators in the lake had to be reptiles, as pound for pound, reptiles consume a fraction the amount of nourishment of mammals. Ten percent, in fact, so that the amount of food a one tonne mammal requires each month, will support a one tonne reptile for one year. I mention this for the sake of those who are not mathematical wizards.

Yet Basilosaurus, the walking whale, is definitely a mammal, and it is definitely alive and well, in Okanagen Lake. Yet it is not the lake, by itself, which supports that whale. My mistake was in thinking of the lake in isolation. In fact, it is part of a large ecosystem, which includes the caves, the streams which flow into the lake, the adjacent meadows, swamps and forested areas. It is the Okanagan ecosystem which supports these huge whales. Further, as top predators, Basilosaurus is key to the health of the whole ecosystem.

That solves one mystery, but in no way changes the fact that the huge prehistoric swimming reptiles must still be alive. As they are clearly not in the fresh water lakes, it stands to reason that they must be in the salt water oceans. Now it is simply a matter of locating them.

With that in mind, some of the techniques I suggested earlier, in looking for Ogopogo, can now be applied to the coastal regions. As these animals are reptiles, they have to come to the surface on a regular basis, in order to breath. So that rules out the deep blue sea. Yet there are few reports, of which I am aware, from sailors, concerning these huge animals. This suggests that the animals are also nocturnal, perhaps also spending the day light hours in caves.

Assuming that is the case, then it is also reasonable to assume that they must be located close to land. The males may have also evolved the same method of display, as that of their flying brethren, the pterosaurs. In other words, these swimming reptiles may be able to glow. If so, then that explains the under water lights, which have so mystified people.

In fact, there are numerous reports of under water lights. The trouble is that the details, which are of such vital importance, are quite scarce.

Among those few rare gems, are the fact that these lights have been spotted "off the coast of California", as well as "close to the Soloman Islands". That is not much to go on, but it is a place to start.

It stands to reason that the caves, where they find shelter, must be on the coast, close to the location of the lights. It further stands to reason that these animals, as reptiles, lay their eggs in the sand of the beach. Equally without doubt, the flying predatory birds, referred to as raptors, birds of prey, are also well aware of the location of this nesting site, and of the time of the year of the hatching of the eggs. After all, the survival of all predators depends upon their being able to take advantage of all food sources.

We can use this to our advantage. Rather than looking for the nesting ground of these reptiles, we can instead look for a gathering of raptors, next to the beach. As they sit on the branches of trees, in anticipation of a feast, and there are great flocks of them, they are east to spot.

Of course, the coast of California is vast, and it would be nice to narrow our search area, but it may not be necessary. It is quite possible that these animals are widespread. My suggestion is that the newly created Councils, or at least those which have taken shape close to the coast, should assign teams of common people to investigate. Each team should be assigned a certain stretch of the coast. Their assignment should be to determine if anyone living along that stretch of the coast has seen any such gathering of raptors. As such coastal areas tend to be densely populated, it is very likely that the people who live there, have noticed such gatherings.

I must stress that this quest, for pre historic animals which the scientist claim to be extinct, is part of the class struggle, and not a substitute for that struggle. Working people are concerned with more than wages, living and working conditions. Mind you, those are important in their own right, and the gaining of such reforms are a by product of the revolutionary movement. In fact, they tend to strengthen and further the revolutionary motion. That is a fact.

It is also a fact that the proof of the existence of various huge species, by the members of the working class, can also be viewed as a reform. After all, they are part of our heritage, a heritage which has been stolen from us, by the capitalists. To restore our heritage, by the same members of the working class, will also have the effect of raising the morale of the working people. It will dramatically increase their self confidence.

We can think of this as valuable training towards the Dictatorship of the Proletariat, because that is precisely the case. After the revolution, after the capitalist are overthrown, and the state apparatus is smashed, numerous workers will have to be placed in positions of authority, in order to crush the "desperate and determined" resistance of the capitalists, as they try to restore their "paradise lost". The training those worker receive now, will prove to be most valuable.

At the same time, this will assist the members of the Councils, in their efforts to determine the suitability of certain workers for key positions, especially at the time of the insurrection. Bear in mind that at the time of the Russian Revolution of October, 1917, the insurrection was not only successful, but also almost bloodless. There was a reason for that. It was well organized. Only the most determined, steadfast workers were placed in key positions, at the time of the insurrection. Those who tended to vacillate, or could even be expected to vacillate, were purged, before the insurrection. This may sound harsh, but only because it is harsh. The time of the insurrection is no time to be sentimental. It is the time to be completely audacious. Any weakness can prove to be fatal. The "defensive is the death of any insurrection"! At such a time, the slogan must be, Victory or Death!

To the Councils, I can only say that the revolution may erupt into full scale civil war, at any time. Such an insurrection has a far greater chance of success, if it is carefully planned, not spontaneous. Proper leadership is critical. This calls for a proper Communist Party, one which calls for the Dictatorship of the Proletariat. That is where the leaders of the Councils come into play. Now is the time to form a proper Communist Party, Dictatorship of the Proletariat. Plan the insurrection, using the working people who have proven themselves in the quest for these animals.

The success of the revolution may well depend upon this.

CHAPTER 8

WORKERS: ASSIST IN LOCATING THE BIG THREE!

There are three huge species of animals which are responsible for numerous legends. Even though the scientists insist they are extinct, they are very much alive, and each is located in North America. They are absolutely not extinct, and with the help of working people, we can quite easily prove this.

As for the skeptical, astute reader, who is wondering just how this is possible, may I refer you to a speech given by Lenin, in October of 1920, three years after the socialist October revolution, to a gathering of the Young Communist League:" It was the declared aim of the old type of school to produce men with an all-round education, to teach the sciences in general. We know that this was utterly false since the whole of society was based and maintained on the division of people into classes, into exploiters and oppressed. Since they were thoroughly imbued with the class spirit, the old schools naturally gave knowledge only to the children of the bourgeoisie. Every word was falsified in the interests of the bourgeoisie. In these schools, the younger generation of workers and peasants were not so much educated as drilled in such a way as to be useful servants of the bourgeoisie, able to create profits for it without disturbing its peace and leisure."

Of course, Lenin was referring to the bourgeois schools, the schools under capitalism. Those are the very schools with which we are currently cursed. As a result of this, those who graduate from these schools, including those who have earned degrees in science, are truly" useful servants of the bourgeoisie", those who know how to" create profits", without" disturbing its peace and leisure".

Without doubt, the act of locating these three species of animals, will certainly" disturb" the" peace and leisure" of the bourgeoisie, the billionaires. It may even interfere with their profit! All the more reason to prove they exist!

The animal which has given rise to the most legends, is the flying reptiles, technically referred to as pterosaurs, more commonly referred to as pterodactyls. In most parts of the world, such as Asia, Africa and Europe, they are referred to as Dragons. Here in North America, there are numerous local names. The more common local names include Thunder Bird, Devil Bird, Satan Bird, Demon Bird and Jersey Devil. These animals are nocturnal, coming out of the caves, in the mountains, mainly after sundown, and hunting in darkness. They prey upon people, as well as livestock. It is only natural that common people should get the idea that these animals are demons!

Of course, they are not demons, but the beliefs of all common people must be respected. It is also absolutely essential that we prove that they exist, if for no other reason than to give warning to people. The fact is that they are predators, and as such, have a very keen sense of smell. Further, they are far more likely to attack, at the smell of blood. For that reason, they consistently prey upon girls of child bearing age. They also prey upon children, as they are easier to pick up and carry away.

This brings me to the best method of proving their existence. Bear in mind that they are famous for releasing a" cloud of smoke", so called, which is not smoke at all, but it is toxic. In fact, it is so strong, it is capable of killing horses and cattle. We know this, for a fact, because that is precisely the manner is which they kill those animals. Then they rip the flesh from the face, as well as the genitals, and frequently tear out the large intestine.

Then in the morning, the owner of these animals stumbles upon these carcasses, furious and bewildered. Such owners generally call the police, which does no good, as it is not a police matter. After all, no laws have been broken. Humans were not involved in these killings. There is no law against wild animals killing livestock! If anything, it is a matter for the game warden. Yet it is rare that the game warden is even notified.

This brings me to the method we can use to prove that these predators exist. First, as soon as possible after the animal is killed, or at least within twenty-four hours, draw out a sample of blood from the carcass. Send this sample to the lab, to be analyzed. The technicians in the lab will in turn determine the poison gas that was used to kill the animal. I am certainly curious. Second, at the same time, swab around the wound sites, and send those swabs also to the lab, for a DNA analysis. No doubt the lab will determine that the wounds were inflicted by a reptile, one which is not known to science. The third thing which has to be done is a bit more complicated, and requires a little explanation.

Close to the village in which I live, that of Tsay Keh Dene, there is a range of mountains, which the locals refer to as Buffalo Head. These mountains are quite distinctive, in that many of them have a very steep, or vertical face, commonly referred to as a cliff, or bluff, while the top of the mountain is horizontal, or flat. I refer to these as perpendicular mountains. It is such mountains that contain the nesting ground of the pterosaurs.

Thirty years ago, a logging road was built into that area, referred to as the Ten Thousand Road. The caves in those mountains, which are the nesting grounds of the pterosaurs, open up onto that road. Now it is a little matter of accessing those caves and placing cameras near the openings. That is not as easy as it sounds, as after the area was logged out, that road was deactivated. The point being that in order to access that nesting ground, the Ten Thousand Road will first have to be activated.

Of course that requires government permits, as well as money, in order to hire the proper equipment. Yet once we establish the fact that a great deal of livestock is being killed by poison gas, released from an animal which is not known to science, then no doubt sufficient pressure can be brought to bear upon the officials. As that is the case, in due time, we can expect them to become sweetly reasonable.

The second animal is also the source of a great many legends. It is nothing other than a separate species of human, technically referred to as Gigantopithecus, commonly called "Sasquatch" or "Bigfoot". The Indigenous people refer to them as "Stink People", a name which is something less than flattering. I refer to them as Giants, as they are huge, and the name is not derogatory. It is essential that we prove they exist, in order to pass laws to protect them. As they are a separate species of human, they have the right to live their lives as they see fit. We have no right to interfere with them. We certainly have no right to hunt them!

These Giants are nomads, members of a hunting-gathering society, so that they are frequently on the move, avoiding us. They know what a vicious bunch we are! Yet they also know that the Indigenous People are more tolerant. For that reason, on the Pacific Coast, the Giants frequently go onto the beaches, on the Reserves, after sundown. The Indigenous People respect the Giants, just as the Giants respect the Indigenous People. That is the way it should be!

That is also the place we can establish contact with the Giants. Rather than hunt them, we want to attract them. Allow them to come to us!

It is all a matter of securing the cooperation of the Elders who live on the Reserves. We must explain to them that we merely want to meet the Giants, in order to prove that they exist. Assuming we can get their cooperation, then it is a matter of putting out gifts for the Giants, on the beach, just before sundown. Bear in mind that as they are human, they love the same things we love. That includes food such as vegetables, fruit, meat, fish and cheese, raw as well as cooked. Also, may I suggest steel mirrors, as they too want to know what they look like. Then there are cosmetics, in order to enhance their appearance. No doubt, they have baskets, but as they tend to be rather flimsy, more sturdy burlap bags will be appreciated.

I can only stress the fact that we want to attract them. No doubt, they will respond to our advances. Give them time, as they have no reason to trust us! Yet they trust the Indigenous People, so perhaps the breakthrough will take place somewhat sooner, rather than later.

The third item on our hit parade is also a source of numerous legends. In North America, it is most commonly referred to as Ogopogo, as it is frequently seen in Okanagan Lake. No doubt, that is a rather clever play on words. Mind you, the animal in Lake Champlain is called Champie. Also very clever! Then there is the animal in Loch Ness, called Nessie, if you can believe it. Very likely the same animal, although that remains to be determined.

Remarkably enough, locating this animal is simplicity itself. For the longest time, I was wondering how a fresh water lake, even a very large lake, could support a population of huge predators. I knew I was missing something, and fortunately, it was recently brought to my attention. This predator has legs!

I am now convinced that Ogopogo is in fact basilosaurus, a species of whale, one which is thought to be extinct. It is not extinct. It is a walking whale! It is a predator, but not a carnivore. It consumes vegetation as well as flesh, which makes it an omnivore. It is this varied diet which enables it to exist in fresh water lakes. It also makes it easy to locate.

This animal is nocturnal, so that it spends most of the daylight hours inside caves. The people who are searching for this animal are focused mainly upon the water, during the day light hours, so of course they have come up empty handed.

It is only after sundown that it comes out of the water and grazes on vegetation, in meadows adjacent to the lake. No doubt, it mainly eats grass, so that the meadow appears to have been mowed.

These" grazed meadows" are the key to locating this animal. May I suggest sending a drone over the banks of the lake, looking for a meadow which appears to have been mowed. That is the grazing ground of basilosaurus. Then place a camera, perhaps even a trail camera, upon a tree at the edge of that meadow. It is very likely that the animal will be seen in the moon light. Mission accomplished!

These are quite simple tasks, yet also revolutionary. The current ruling class, the billionaires, the bourgeoisie, do not want the members of the working class, the proletariat, being made aware of the existence of these huge species! That could disturb their "peace and leisure"! God forbid! Worse, it could even threaten their profits! People could even have the gall to demand that the polluted lakes and other areas be cleaned up! The nerve of some people! Better to continue to deny the existence of these animals! Listen to the scientists! They are well trained, loyal,"useful servants"!

The scientists are also a pack of liars, as they cannot possibly be so stupid, so completely incompetent, as to be unaware of the existence of these huge animals!

This has to change, and quickly. The capitalists, the billionaires, the bourgeoisie, must be overthrown, and it is up to the working class, the proletariat, to accomplish this. The revolutionary motion is currently sweeping the country. Part of this revolution must consist of proving the existence of these huge species. In the process, the scientists will be proven to be the loyal servants of the bourgeoisie, which is precisely the case.

In this manner, the working people will gain valuable experience in the class struggle. It will help to strengthen and further the revolutionary motion, as workers gain more confidence.

It may help to think of this as preparation for the Dictatorship Of the Proletariat, because that is precisely the case. After the revolution, workers will be placed in positions of authority, within the new socialist government. Any training workers receive now, will prove to be of great value, after the revolution. The act of taking part in a series of major scientific break through, can be considered to be part of that training.

My hope is that cells of workers will come together, led by Councils, and cooperate in proving the existence of these magnificent animals. These cells can also form the nucleus of revolutionary bodies. Success will no doubt attract others, so that the cells are bound to grow. Let the slogan be:

Prepare For the Dictatorship of the Proletariat

CHAPTER 9

PREPARE FOR THE DICTATORSHIP OF THE PROLETARIAT

The fact is that I have recently identified the animal in Oganagan Lake, "Ogopogo", as Basilosaurus, a whale with legs, a "walking whale". This opens up various possibilities, opportunities for people who are taking part in the revolutionary motion, to become ever more politically active. In fact, they can take part in a major scientific break through, one of several. The political struggle should not, and must not, be confined to demanding higher wages and better working conditions.

Before I proceed, I should mention that I make a point of writing in a very popular manner, with working people in mind. At least, that is the way in which they refer to themselves. In fact, such people are members of the working class, proletarians, although few of them are aware of this. So now my goal is to motivate working people to become politically active, while at the same time, bring to them the awareness of the existence of classes, including the fact that they are members of a working class.

This involves the explanation of scientific terms. Those who are already class conscious, may find this to be tiresome, but it cannot be helped. Also, for the purposes of this article, I use the term "capitalist" to refer to the super rich, the billionaires, technically referred to as the bourgeoisie. It is important to distinguish them from the small time capitalist, the small business owner, technically referred to as the petty bourgeois. The working class has no quarrel with them, just as we have no quarrel with the professional, salaried employees. In fact, they are the natural, desirable allies of the proletariat. I am also careful to avoid the personal names of most people, to whom I refer in my articles, as I do not want to face a law suit. Now to the heart of the matter.

The current crisis in capitalism has led to some interesting developments, which are to be expected, as they have happened so often in the past. The revolutionary motion has had the unintentional consequence of highlighting two separate tendencies in International Marxism, otherwise known as Communism, formerly known as Social Democracy or Bolshevism. On the one hand, there are the social chauvinists, who maintain that Communism must change from a party of social revolution, into a democratic party of social reform. They reject the idea of *scientific socialism,* especially the Marxist theory of the *Dictatorship of the Proletariat!* The *touchstone* of a true Marxist is absolutely rejected! They think that socialism and liberalism are the same thing! The *theory of the class struggle is rejected,* on the grounds that it cannot be applied to a democratic society, which implies "majority rule", and other such nonsense. Yet all too many of them insist on referring to themselves as Marxists!

Lenin dealt with such people extensively, and in his book What Is To Be Done?, referred to them as Economists, or Mensheviks. They consistently choose the path of conciliation, rather than the path of struggle. He refers to such conciliators as those who are "in the swamp". They persist to this day, and in fact there is no shortage of them. They insist that there is no need for a revolutionary theory, despite the fact that Engels stated, most emphatically, that *"without a revolutionary theory, there can be no revolutionary motion"!* In fact, Engels recognized three great struggles of Communism, *political, economic and theoretical!*

This brings me to the subject of Ogopogo, which is really basilosaurus, a whale with legs, a "walking whale". Most working people are fascinated with the legend of Ogopogo, and no doubt will be anxious to take part in proving the fact that it exists, or at least the fact that the legend is based on a real animal. At the same time, it will help to boost their self confidence, as so many working class people have placed scientists "on a pedestal". To take part in proving that the scientists are mistaken, will have the effect of empowering the workers. Their self confidence will increase dramatically. It will inspire them to challenge other figures of authority, including the billionaires. The precise method employed, in this noble endeavour, is critical.

Having said that, we can now consider the "Councils", a creation of the revolutionary movement. These Council members are composed of leaders of the working class, and such people plot a course of action for the working class. In the city of Seattle, they actually set up a Zone, and declared it to be Autonomous. They very quickly learned, just as the workers who took part in the Occupy Movement learned, that such Zones are not allowed. The capitalists consider such Zones to be a threat to their authority, as indeed they are. The Zone in Seattle, referred to as the Capitol Hill Autonomous Zone, was quickly crushed. Yet the Councils remain. May I suggest that these Councils assist in taking part in proving the existence of these huge animals, and in the process, train countless working people.

Incidentally, many people may not be aware of the fact that Soviet is a Russian word, which means Council. They are spontaneous creations of the working class, and first appeared in Russia, in 1905, at the time of the first Russian revolution. As they eventually gave rise to the Soviet Union, they are not to be under estimated. It is not a coincidence that these Councils have also taken shape here, at this time. We too, are on the eve of a revolution. We have got to be prepared, and there is no time to waste. As for those who are skeptical, thinking perhaps that a revolution could not possibly happen here, feel free to face the facts. Bear in mind that the interests of the two classes, the proletariat and the bourgeoisie, are diametrically opposed. The wealth of the capitalists comes at the expense of the workers. The more wealth the capitalists gather, the more impoverished the workers. Yet according to the bourgeois economists, since the start of the pandemic, less than two years ago, the *wealth of the billionaires has doubled!*

These same bourgeois economists also report that the billionaires pay little or no taxes. Among other tricks of the trade, they may have "no income". They may merely "borrow" money, from a company they own, as a means of supporting their lavish life style. Tax free. There are numerous other ways to avoid paying taxes, all of which are *perfectly legal!*

There are now a number of billionaires who are worth tens, and even hundreds of billions! It is not enough! Each and every one of them wants to be the first *Trillionaire!* In other words, they want to possess the wealth of a thousand billion!

As one of these multi billionaires explained, in response to the suggestion that he should pay his "fair share" of taxes: "Eventually, they run out of other peoples money, and then they come after yours". The billionaires see themselves as *victims!* Another billionaire stated it more prosaically, in response to the suggestion that he pay taxes, with a possible attempt to sound profound: "It is better to get humanity to Mars, and preserve the light of consciousness". Rather than pay taxes, in order to provide housing for the homeless, medical care for those whom so desperately need it, food for the hungry, repairs to the roads and bridges, among a great many other things, he considers it more important to "preserve the light of consciousness" by" sending humanity fo Mars". To think that the social chauninists would have us believe that such people are about to "*submit to the will of the majority*"! Not likely!

The point is that the situation is truly revolutionary, as there is a limit to that which people are about to tolerate. With that in mind, may I suggest that some of these newly created Councils, especially those which are located close to large fresh water lakes, take steps to verify the existence of these walking whales. It is not difficult, and the results will be immediate and dramatic. It is not just Okanagan Lake that is home to this animal. It is also in Lake Champlain, and is known to locals as the Champ. In Lake Erie, it is known as Bessie. Very likely, each and every one of the Great Lakes is home to this animal. For that matter, it is very likely that the Loch Ness monster is none other than Basilosaurus. There have been reports of this animal located in other large lakes in Europe. Councils in those European countries are of course free to follow suit. This I recommend.

In any case, all Councils should mobilize as many people as possible. It is to be expected that sports enthusiasts, such as members of rod and gun clubs, will be especially interested, as well as students of science. Yet as so many working people are interested in these legends, the response should be impressive. Then it is a matter of breaking them up into teams, perhaps with a mixture of young and old. Impress upon all that there is an element of danger to this, so that reasonable precautions are expected. Then each team should be assigned a sector of the lake, along with a map of the lake, as well as a drone and various trail cameras.

The idea is that each team is responsible for sending their drone over their sector of the lake, and plotting out the meadows located on the perimeter of the lake. Those meadows are the grazing ground of basilosaurus. Such a task could well take all day. The following day, each team can go to the meadows which they have located, and attach a trail camera to a tree, on the edge of the meadow. At least one member of the team should carry a high powered rifle. The idea is to protect the team, in case of an attack by a predator.

It is very likely that these trail cameras will be able to detect basilosaurus, in the moon light. After all, it is twenty meters long, and weighs fifteen tonnes, so it is rather difficult to miss. Yet that also means that the cameras will have to be checked, on a regular basis.

Of course, it is important to prove the existence of this animal. Yet the more important aspect of this little exercise is to train people to work together, as a team, independently. At the same time, their performance can be quietly evaluated. This is valuable training for the revolution, and the subsequent Dictatorship of the Proletariat. Yet first comes the Insurrection, the day of reckoning. On that day, various key locations, across the country, will have to be secured by the revolutionary proletariat. It is not enough to take possession of the Vipers Nest, in the capital of Washington, D.C. Mind you, that little task will likely not be terribly difficult, as the events of January 6 have revealed. But then the capitalists have also noticed the weakness in their defences. So they have responded by building a wire fence around the buildings. Their childish faith in fences is somewhat touching, even if it is quite pathetic.

As I have documented in a previous article, on the day of the Insurrection, it will also be necessary to shut down the railroads, bridge, tunnels, airports and sea ports, as well as communications networks. This is to say that all across the country, numerous groups of revolutionaries, working people, will have to take action, under the direction of a local Council. Each local Council, in turn, must work under the supervision of a national authority. We will go into that detail later, in this article. Each and every one of these groups must be resolute. If even one group fails to carry out its assignment, the fate of the Insurrection could be in jeopardy.

The point is that the quest for these huge animals could serve as a training exercise. May I suggest that as many Councils as possible, assign as many groups of workers as possible, in an attempt to locate these animals. The main thing is to evaluate the workers, to determine those who are the most resolute. At the same time, take note of those who are somewhat indifferent. This is not to say that we are trying to judge people. It is to say that on the day of the Insurrection, we must use only those who are completely resolute. We have no use, on that day, for those who are likely to waver. There are other animals which have yet to be proven to exist. One of these is the Giants, or Gigantopithecus, otherwise known as Sasquatch or Bigfoot, and it too, is quite easy to locate. They are on the Reserves along the Pacific Coast, and it requires the cooperation of the Indigenous people, in order to make contact with these people. Once that happens, working people can prove the existence of another huge species. This will also provide the Councils with more valuable information, concerning the commitment and determination of various workers.

Then there are the pterosaurs, commonly referred to as dragons or thunder birds, although the list of local names is endless, and locating them is a bit more of a challenge. May I suggest that people pay attention to any reports of cattle or horses that are found dead in the morning, terribly mutilated, but with no visible sign of blood. Then it is a matter of drawing out a sample of blood, as well as swabbing around the wound sites. Both should be then rushed to a lab for analysis. The lab will then determine the poison gas that was used to kill the animal, and further, that the DNA is that of a reptile, one not known to science. Then it is a matter of getting the permits to open up the old ten thousand logging road, which leads to the nesting ground of the pterodactyls. Cameras can then be placed at the entrance to the caves.

These are very simple tasks, which could well be carried out by the scientists. They are not carrying them out, because the last thing they want to do is to "disturb the peace and tranquillity of the capitalists". That is the first thing we want to do! At the same time, workers will receive valuable training, and members of the Councils will get a chance to determine the suitability of those workers, to be put to work, at the time of the Insurrection.

In particular, proof of the existence of walking whales will cause a major uproar. There will be an immediate call for the lakes, especially the Great Lakes, to be cleaned up. As they are severely polluted, mainly as a result of industrial run off, people will demand that the factories clean up their own mess. That is the last thing the capitalists want to hear! Perhaps if the politicians would spend less time squawking about climate change, and focus more on cleaning up our environment, then we would live in a far better world.

It is entirely possible that the revolution has already started. Recently, a major airline cancelled hundreds of flights, over a period of several days. The press reports are rather vague, but there are references to people who refused to work, including pilots and air traffic controllers. Assuming that such people are members of a union, then such "walkouts" are referred to as "wild cat strikes". This is a reference to a strike that has not been authorized by the union leaders. If that is the case, then it is an indication of the strength of the revolutionary motion.

Bear in mind that most revolutions start with strikes in the transportation industry. That includes railroads, airlines and shipping lines. As the capitalists are complaining about delays in shipping, it is entirely possible that the workers are engaging in "slowdowns". That is not exactly a strike, but very often a prelude to a strike. Such slowdowns take place as a result of deep worker dissatisfaction. As most union leaders are "in the pocket" of the capitalists, such dissatisfaction is completely understandable.

Strikes are one thing and Insurrection is something else entirely! At some point, the working people who are taking part in the revolution, have to seize political power! That calls for an Insurrection! That is not something to be taken lightly! The vast majority of the workers, or at least of the most advanced workers, must be prepared to overthrow the capitalists, smash the existing state apparatus, and establish the Dictatorship of the Proletariat.

The key word here is "prepare"! The most advanced workers must become class conscious, aware of the existence of classes. They must also become aware of the revolutionary theories of Marx and Lenin. The necessity of smashing the existing state machine, and replacing it with the Dictatorship of the Proletariat, must be stressed. Another part of that preparation involves becoming organized, working together as a team, as part of a large army. The experience that working people are about to gain, in proving the existence of these huge animals, will prove to be valuable training.

Bear in mind that in a similar situation, that of Russia in 1917, Lenin returned from exile in April, after the Tsar had been overthrown, and a democratic republic had been established. Yet Lenin did not immediately call for an Insurrection. In fact, a possible uprising in July of that year was aborted, as it was thought that the working class was not properly prepared.

Those days have gone down in history as the "Revolutionary July Days". I mention this because it is so important. Lenin called off a possible Insurrection, at that time, as it was clear to him that the working class, or at least the most advanced strata of the proletariat, had not fully embraced the Dictatorship of the Proletariat. This is another way of saying that the Russian proletariat was, at that time, not sufficiently class conscious. It was up to the Communists to raise the level of awareness, of the advanced workers, to that of the level of Communists. This they then managed, which in turn made possible the Insurrection, several months later, on October 25, old stye calendar, or November 7, now style calendar.

The American proletariat of today is even less class conscious than the Russian proletariat of 1917, through no fault of their own. The conditions of life, of the proletariat, do not lead to the awareness of itself, as a class. This awareness must be brought to it, from an outside source. That is the duty of middle class intellectuals.

We clearly have our work cut out for us, but that is no reason for despair. Most working people are literate, and most of them have access to digital devices. The task of raising their level of consciousness is far easier for us, than it was for the Russian Communists of 1917. Among other things, we have the internet, and we would be fools not to take advantage of it. It just means making popular literature available for the workers, very popular literature for the less advanced, but by no means vulgar. Feel free to use sports metaphors, and avoid the use of the word "backward", when referring to workers. So many workers may consider this to mean "stupid", and the last thing we want to do is offend any member of the working class. We want to flood social media with such literature. No doubt, leaders will emerge.

We would do well to bear in mind that working class people are avid readers. They pay strict attention to the news, so that in the literature, be sure to use current events. I mention the Russian revolution, as that is the revolution which most closely resembles our own. Granted, there are considerable differences. In the case of Russia, the existence of the nobility, landlords and peasants, each with their own class interests, complicated that situation. Our situation is much simpler, in that we have the bourgeoisie and the proletariat. All other classes have been all but wiped out.

This is to say that we live in a highly industrialized, or "cultured" country, as opposed to an under developed, or "petty bourgeois" country. I mention this for the sake of those who are just now becoming politically active. It also means that starting a revolution is this country is much more difficult, as the bourgeois ideology is so deeply entrenched. Yet carrying the revolution through, after the Insurrection, is much easier. It is also a fact that there is an urgent need for a true Communist Party, one which calls for the Dictatorship of the Proletariat. After all, people need leaders. Workers can only do so much! It is very likely that many members of the newly created Councils are conscious people, well educated, either current or former members of the middle class. Such people tend to be well aware of the revolutionary theories of Marx and Lenin. Precisely the sort of people we need to create a true Communist Party!

To such people, may I suggest that the creation of Councils is excellent, a step on the right direction, but only a step. Half measures get us nowhere! The next step involves getting together with other conscious people, Marxists, possibly from other Councils, and creating a true Communist Party.

No doubt, all members of the Councils are "Leftists", and consider themselves to be socialists, or are at least sympathetic to socialism. Equally without doubt, many of them are well aware of the revolutionary theories of Marx and Lenin. May I suggest that now is the time to apply those theories to a revolutionary situation. This is to say that a true Communist Party, one which calls for the Dictatorship of the Proletariat, is ungently needed.

The creation of such a Party may not be terribly difficult. No doubt, the various Councils are in touch with each other, so that the Communists on each Council can get together, not necessarily in person, and create a Party. The internet program of Skype comes to mind. Feel free to avoid certain words, as the government computers are programmed to "flag" such conversations. Do not make anything easy for the government agents! Bear in mind that as the pedophiles are able to use the internet, while escaping detection, so can we!

This is not to say that the Communist Party should take the place of the Councils. On the contrary, the Party should work as closely as possible with the Councils, as well as with the trade unions, cooperative societies and sports clubs. That same Party can be the national authority, and at the time of the Insurrection, coordinate the activities of the local Councils.

There is not time to lose! Either the revolution will led by Communists, or it will be led by reactionaries, such as Trump! Your choice! For the moment, the slogan of all conscious people must be:

Prepare For the Dictatorship of the Proletariat!

CHAPTER 10

In Quest of Pre Historic Swimming Reptiles

As is well known, the scientists maintain that the "dinosaurs" went extinct sixty-five million years ago. They also maintain that the "great sea lizards and the snake necked plesiosaurs were also dying out", at that time. It is no secret that I consider this to be nothing other than a scientific fairy tale. I have documented that most of the animals which have been classified as dinosaurs were nothing other than birds, while others, such as the flying pterosaurs, were reptiles. I also maintain that the "great sea lizards and snake necked plesiosaurs" are still alive, as are the pterosaurs.

With that in mind, I suspected that the animal in Okanagan Lake, the so called "Ogopogo", was nothing other than one of those swimming reptiles. Yet I was mistaken, as I have recently determined that the animal in Okanagan Lake, is nothing other than basilosaurus, a huge whale, a whale with legs, a walking whale.

As I have documented in a previous article, this makes the task of proving the existence of this animal much easier. Basilosaurus is a predator, but not a carnivore. It is an omnivore, which is to say that it consumes flesh and vegetation. It is also nocturnal, in that it spends most of the daylight hours in caves, which are accessible from the water. I should add that common people refer to these caves as "underwater caves". They are not under water, but it is perfectly acceptable for common people to refer to those caves in that manner. It is up to us, professional people, to determine the fact that those caves are above the water line, so that the caves provide the animal with shelter, warmth and a place to rest.

Then after sun down, the animal comes out of the water and consumes vegetation, mainly grass, no doubt. That makes the task of proving the existence of the animal simplicity itself. It is simply a matter of setting up trail cameras, on the edge of the meadows, adjacent to the lake. In the moon light, we should be able to get some fine pictures of the animal. As it is the largest of all land dwelling animals, it should not be difficult to locate.

My mistake was instructive so that possibly others can learn from this. Common sense told me that a fresh water lake could not possibly support a population of huge predatory swimming mammals. I still maintain that I was correct. So I assumed that the predators in the lake had to be reptiles, as pound for pound, reptiles consume a fraction the amount of nourishment of mammals. Ten percent, in fact, so that the amount of food a one tonne mammal requires each month, will support a one tonne reptile for one year. I mention this for the sake of those who are not mathematical wizards.

Yet basilosaurus, the walking whale, is definitely a mammal, and it is definitely alive and well, in Okanagen Lake. Yet it is not the lake, by itself, which supports that whale. My mistake was in thinking of the lake in isolation. In fact, it is part of a large ecosystem, which includes the caves, the streams which flow into the lake, the adjacent meadows, swamps and forested areas. It is the Okanagan ecosystem which supports these huge whales. Further, as top predators, basilosaurus is key to the health of the whole ecosystem.

That solves one mystery, but in no way changes the fact that the huge prehistoric swimming reptiles must still be alive. As they are clearly not in the fresh water lakes, it stands to reason that they must be in the salt water oceans. Now it is simply a matter of locating them.

With that in mind, some of the techniques I suggested earlier, in looking for Ogopogo, can now be applied to the coastal regions. As these animals are reptiles, they have to come to the surface on a regular basis, in order to breathe. So that rules out the deep blue sea. Yet there are few reports, of which I am aware, from sailors, concerning these huge animals. This suggests that the animals are also nocturnal, perhaps also spending the day light hours in caves.

Assuming that is the case, then it is also reasonable to assume that they must be located close to land. The males may have also evolved the same method of display, as that of their flying brethren, the pterosaurs. In other words, these swimming reptiles may be able to glow. If so, then that explains the underwater lights, which have so mystified people.

In fact, there are numerous reports of underwater lights. The trouble is that the details, which are of such vital importance, are quite scarce. Among those few rare gems, are the fact that these lights have been spotted "off the coast of California", as well as "close to the Soloman Islands". That is not much to go on, but it is a place to start.

It stands to reason that the caves, where they find shelter, must be on the coast, close to the location of the lights. It further stands to reason that these animals, as reptiles, lay their eggs in the sand of the beach. Equally without doubt, the flying predatory birds, referred to as raptors, birds of prey, are also well aware of the location of this nesting site, and of the time of the year of the hatching of the eggs. After all, the survival of all predators depends upon their being able to take advantage of all food sources.

We can use this to our advantage. Rather than looking for the nesting ground of these reptiles, we can instead look for a gathering of raptors, next to the beach. As they sit on the branches of trees, in anticipation of a feast, and there are great flocks of them, they are easy to spot.

Of course, the coast of California is vast, and it would be nice to narrow our search area, but it may not be necessary. It is quite possible that these animals are widespread.

My suggestion is that the newly created Councils, or at least those which have taken shape close to the coast, should assign teams of common people to investigate. Each team should be assigned a certain stretch of the coast. Their assignment should be to determine if anyone living along that stretch of the coast has seen any such gathering of raptors. As such coastal areas tend to be densely populated, it is very likely that the people who live there, have noticed such gatherings.

I must stress that this quest, for pre historic animals which the scientist claim to be extinct, is part of the class struggle, and not a substitute for that struggle. Working people are concerned with more than wages, living and working conditions. Mind you, those are important in their own right, and the gaining of such reforms are a by product of the revolutionary movement. In fact, they tend to strengthen and further the revolutionary motion. That is a fact.

It is also a fact that the proof of the existence of various huge species, by the members of the working class, can also be viewed as a reform. After all, they are part of our heritage, a heritage which has been stolen from us, by the capitalists. To restore our heritage, by the same members of the working class, will also have the effect of raising the morale of the working people. It will dramatically increase their self confidence.

We can think of this as valuable training towards the Dictatorship of the Proletariat, because that is precisely the case. After the revolution, after the capitalist are overthrown, and the state apparatus is smashed, numerous workers will have to be placed in positions of authority, in order to crush the "desperate and determined" resistance of the capitalists, as they try to restore their "paradise lost". The training those worker receive now, will prove to be most valuable.

At the same time, this will assist the members of the Councils, in their efforts to determine the suitability of certain workers for key positions, especially at the time of the Insurrection. Bear in mind that at the time of the Russian Revolution of October, 1917, the Insurrection was not only successful, but also almost bloodless. There was a reason for that. It was well organized. Only the most determined, steadfast workers were placed in key positions, at the time of the Insurrection. Those who tended to vacillate, or could even be expected to vacillate, were purged, before the Insurrection. This may sound harsh, but only because it is harsh. The time of the Insurrection is no time to be sentimental. It is the time to be completely audacious. Any weakness can prove to be fatal. The "defensive is the death of any insurrection"! At such a time, the slogan must be, Victory or Death!

To the Councils, I can only say that the revolution may erupt into full scale civil war, at any time. Such an Insurrection has a far greater chance of success, if it is carefully planned, not spontaneous. Proper leadership is critical. This calls for a proper Communist Party, one which calls for the Dictatorship of the Proletariat. That is where the leaders of the

Councils come into play. Now is the time to form a proper Communist Party, Dictatorship of the Proletariat. Plan the Insurrection, using the working people who have proven themselves in the quest for these animals.

The success of the revolution may well depend upon this.

www.ingramcontent.com/pod-product-compliance
Lightning Source LLC
Chambersburg PA
CBHW060251030426
42335CB00014B/1653